Survey of Higher Education Faculty 2025, Supplying Content to AI Models

ISBN 979-8-88517-308-7

Table of Contents

QUESTIONNAIRE 15
Characteristics of the sample 19
INSTITUTIONAL AFFILIATIONS OF THE SURVEY PARTICIPANTS 22
SUMMARY OF MAIN FINDINGS 28
Q1. Do you feel that you have content connected to your research that could be used to train an AI model? Summary of Answers. 28
Q2. Have you submitted your content for inclusion in an AI model? Summary of Answers 29
Q3. To whom have you submitted your content for AI model development? Summary of Answers 30
Q4. To the best of your knowledge, has your content been used by AI without your permission? Summary of Answers 31
Q5. Do you know anyone who feels their content has been used without permission? Summary of Answers 31
Summary of Open-ended Comments About Perception of Use of Academic Contents by AI Model Developers 32
Question 6: Do you believe that your institution should negotiate with AI companies to obtain compensation for the use of faculty content in training AI models? Summary of Answers 32
Question 7: How would you feel if your content were used to train an AI model without your permission? Summary of Answers 33
Overall Summary 34
Across Questions 1–7: 34

Table 1.1 Do you feel that you have articles, data, lab logs, code, blog posts, class records or videos, personally developed educational materials, or any other content connected to your research that could be used to train an AI model? 35
Table 1.2 Do you feel that you have articles, data, lab logs, code, blog posts, class records or videos, personally developed educational materials, or any other content connected to your research that could be used to train an AI model? Broken out by enrollment 35
Table 1.3 Do you feel that you have articles, data, lab logs, code, blog posts, class records or videos, personally developed educational materials, or any other content connected to your research that could be used to train an AI model? Broken out by type of college or Carnegie Class 36
Table 1.4 Do you feel that you have articles, data, lab logs, code, blog posts, class records or videos, personally developed educational materials, or any other content connected to your research that could be used to train an AI model? Broken out by public or private college 37
Table 1.5 Do you feel that you have articles, data, lab logs, code, blog posts, class records or videos, personally developed educational materials, or any other content connected to your research that could be used to train an AI model? Broken out by age of respondent 37

Table 1.6 Do you feel that you have articles, data, lab logs, code, blog posts, class records or videos, personally developed educational materials, or any other content connected to your research that could be used to train an AI model? Broken out by political views 38
Table 1.7 Do you feel that you have articles, data, lab logs, code, blog posts, class records or videos, personally developed educational materials, or any other content connected to your research that could be used to train an AI model? Broken out by annual personal income 39
Table 1.8 Do you feel that you have articles, data, lab logs, code, blog posts, class records or videos, personally developed educational materials, or any other content connected to your research that could be used to train an AI model? Broken out by tenure track 40
Table 1.9 Do you feel that you have articles, data, lab logs, code, blog posts, class records or videos, personally developed educational materials, or any other content connected to your research that could be used to train an AI model? Broken out by gender of respondent 41
Table 1.10 Do you feel that you have articles, data, lab logs, code, blog posts, class records or videos, personally developed educational materials, or any other content connected to your research that could be used to train an AI model? Broken out by race or ethnicity 42
Table 1.11 Do you feel that you have articles, data, lab logs, code, blog posts, class records or videos, personally developed educational materials, or any other content connected to your research that could be used to train an AI model? Broken out by academic field 43
Table 1.12 Do you feel that you have articles, data, lab logs, code, blog posts, class records or videos, personally developed educational materials, or any other content connected to your research that could be used to train an AI model? Broken out by years at present institution 44
Table 2.1 Have you submitted any of your articles, data, lab logs, code, blog posts, class records or videos, personally developed educational materials or any other content connected to your research for inclusion in an AI model? 45
Table 2.2 Have you submitted any of your articles, data, lab logs, code, blog posts, class records or videos, personally developed educational materials or any other content connected to your research for inclusion in an AI model? Broken out by enrollment. 45
Table 2.3 Have you submitted any of your articles, data, lab logs, code, blog posts, class records or videos, personally developed educational materials or any other content connected to your research for inclusion in an AI model? Broken out by type of college or Carnegie Class 46
Table 2.4 Have you submitted any of your articles, data, lab logs, code, blog posts, class records or videos, personally developed educational materials or any other content connected to your research for inclusion in an AI model? Broken out by public or private college 46
Table 2.5 Have you submitted any of your articles, data, lab logs, code, blog posts, class records or videos, personally developed educational materials or any other content connected to your research for inclusion in an AI model? Broken out by age of respondent 47

Table 2.6 Have you submitted any of your articles, data, lab logs, code, blog posts, class records or videos, personally developed educational materials or any other content connected to your research for inclusion in an AI model? Broken out by political views 47
Table 2.7 Have you submitted any of your articles, data, lab logs, code, blog posts, class records or videos, personally developed educational materials or any other content connected to your research for inclusion in an AI model? Broken out by annual personal income 48
Table 2.8 Have you submitted any of your articles, data, lab logs, code, blog posts, class records or videos, personally developed educational materials or any other content connected to your research for inclusion in an AI model? Broken out by tenure track 48
Table 2.9 Have you submitted any of your articles, data, lab logs, code, blog posts, class records or videos, personally developed educational materials or any other content connected to your research for inclusion in an AI model? Broken out by gender of respondent 49
Table 2.10 Have you submitted any of your articles, data, lab logs, code, blog posts, class records or videos, personally developed educational materials or any other content connected to your research for inclusion in an AI model? Broken out by race or ethnicity 49
Table 2.11 Have you submitted any of your articles, data, lab logs, code, blog posts, class records or videos, personally developed educational materials or any other content connected to your research for inclusion in an AI model? Broken out by academic field 50
Table 2.12 Have you submitted any of your articles, data, lab logs, code, blog posts, class records or videos, personally developed educational materials or any other content connected to your research for inclusion in an AI model? Broken out by years at present institution 51
If you have already contributed content to train an AI application briefly describe what you have contributed and, if possible, to what kind of effort. 52
Table 3 To which of the following have you ever contributed your content outlined in the past two questions for the development of an AI model by any of the parties below? (Check all that apply) 56
Table 3.1.1 To which of the following have you ever contributed your content outlined in the past two questions for the development of an AI model by any of the parties below? Private Tech Company Developing an AI Model. 56
Table 3.1.2 To which of the following have you ever contributed your content outlined in the past two questions for the development of an AI model by any of the parties below? Private Tech Company Developing an AI Model. Broken out by enrollment 56
Table 3.1.3 To which of the following have you ever contributed your content outlined in the past two questions for the development of an AI model by any of the parties below? Private Tech Company Developing an AI Model. Broken out by type of college or Carnegie Class 56
Table 3.1.4 To which of the following have you ever contributed your content outlined in the past two questions for the development of an AI model by any of the

parties below? Private Tech Company Developing an AI Model. Broken out by public or private college 57
Table 3.1.5 To which of the following have you ever contributed your content outlined in the past two questions for the development of an AI model by any of the parties below? Private Tech Company Developing an AI Model. Broken out by age of respondent........ 57
Table 3.1.6 To which of the following have you ever contributed your content outlined in the past two questions for the development of an AI model by any of the parties below? Private Tech Company Developing an AI Model. Broken out by political views........ 57
Table 3.1.7 To which of the following have you ever contributed your content outlined in the past two questions for the development of an AI model by any of the parties below? Private Tech Company Developing an AI Model. Broken out by annual personal income 58
Table 3.1.8 To which of the following have you ever contributed your content outlined in the past two questions for the development of an AI model by any of the parties below? Private Tech Company Developing an AI Model. Broken out by tenure track........ 58
Table 3.1.9 To which of the following have you ever contributed your content outlined in the past two questions for the development of an AI model by any of the parties below? Private Tech Company Developing an AI Model. Broken out by gender of respondent 58
Table 3.1.10 To which of the following have you ever contributed your content outlined in the past two questions for the development of an AI model by any of the parties below? Private Tech Company Developing an AI Model. Broken out by race or ethnicity 59
Table 3.1.11 To which of the following have you ever contributed your content outlined in the past two questions for the development of an AI model by any of the parties below? Private Tech Company Developing an AI Model. Broken out by academic field........ 59
Table 3.1.12 To which of the following have you ever contributed your content outlined in the past two questions for the development of an AI model by any of the parties below? Private Tech Company Developing an AI Model. Broken out by years at present institution 60
Table 3.2.1 To which of the following have you ever contributed your content outlined in the past two questions for the development of an AI model by any of the parties below? Journal, Book or Other Content Publisher. 60
Table 3.2.2 To which of the following have you ever contributed your content outlined in the past two questions for the development of an AI model by any of the parties below? Journal, Book or Other Content Publisher. Broken out by enrollment 60
Table 3.2.3 To which of the following have you ever contributed your content outlined in the past two questions for the development of an AI model by any of the parties below? Journal, Book or Other Content Publisher. Broken out by type of college or Carnegie Class 61

Table 3.2.4 To which of the following have you ever contributed your content outlined in the past two questions for the development of an AI model by any of the parties below? Journal, Book or Other Content Publisher. Broken out by public or private college 61
Table 3.2.5 To which of the following have you ever contributed your content outlined in the past two questions for the development of an AI model by any of the parties below? Journal, Book or Other Content Publisher. Broken out by age of respondent........ 61
Table 3.2.6 To which of the following have you ever contributed your content outlined in the past two questions for the development of an AI model by any of the parties below? Journal, Book or Other Content Publisher. Broken out by political views 62
Table 3.2.7 To which of the following have you ever contributed your content outlined in the past two questions for the development of an AI model by any of the parties below? Journal, Book or Other Content Publisher. Broken out by annual personal income 62
Table 3.2.8 To which of the following have you ever contributed your content outlined in the past two questions for the development of an AI model by any of the parties below? Journal, Book or Other Content Publisher. Broken out by tenure track 62
Table 3.2.9 To which of the following have you ever contributed your content outlined in the past two questions for the development of an AI model by any of the parties below? Journal, Book or Other Content Publisher. Broken out by gender of respondent........ 63
Table 3.2.10 To which of the following have you ever contributed your content outlined in the past two questions for the development of an AI model by any of the parties below? Journal, Book or Other Content Publisher. Broken out by race or ethnicity 63
Table 3.2.11 To which of the following have you ever contributed your content outlined in the past two questions for the development of an AI model by any of the parties below? Journal, Book or Other Content Publisher. Broken out by academic field 64
Table 3.2.12 To which of the following have you ever contributed your content outlined in the past two questions for the development of an AI model by any of the parties below? Journal, Book or Other Content Publisher. Broken out by years at present institution 65
Table 3.3.1 To which of the following have you ever contributed your content outlined in the past two questions for the development of an AI model by any of the parties below? Peers in my or Related Field. 65
Table 3.3.2 To which of the following have you ever contributed your content outlined in the past two questions for the development of an AI model by any of the parties below? Peers in my or Related Field. Broken out by enrollment........ 65
Table 3.3.3 To which of the following have you ever contributed your content outlined in the past two questions for the development of an AI model by any of the parties below? Peers in my or Related Field. Broken out by type of college or Carnegie Class........ 66

Table 3.3.4 To which of the following have you ever contributed your content outlined in the past two questions for the development of an AI model by any of the parties below? Peers in my or Related Field. Broken out by public or private college 66
Table 3.3.5 To which of the following have you ever contributed your content outlined in the past two questions for the development of an AI model by any of the parties below? Peers in my or Related Field. Broken out by age of respondent........ 66
Table 3.3.6 To which of the following have you ever contributed your content outlined in the past two questions for the development of an AI model by any of the parties below? Peers in my or Related Field. Broken out by political views 67
Table 3.3.7 To which of the following have you ever contributed your content outlined in the past two questions for the development of an AI model by any of the parties below? Peers in my or Related Field. Broken out by annual personal income 67
Table 3.3.8 To which of the following have you ever contributed your content outlined in the past two questions for the development of an AI model by any of the parties below? Peers in my or Related Field. Broken out by tenure track 67
Table 3.3.9 To which of the following have you ever contributed your content outlined in the past two questions for the development of an AI model by any of the parties below? Peers in my or Related Field. Broken out by gender of respondent. 68
Table 3.3.10 To which of the following have you ever contributed your content outlined in the past two questions for the development of an AI model by any of the parties below? Peers in my or Related Field. Broken out by race or ethnicity........ 68
Table 3.3.11 To which of the following have you ever contributed your content outlined in the past two questions for the development of an AI model by any of the parties below? Peers in my or Related Field. Broken out by academic field 69
Table 3.3.12 To which of the following have you ever contributed your content outlined in the past two questions for the development of an AI model by any of the parties below? Peers in my or Related Field. Broken out by years at present institution........ 70
Table 3.4.1 To which of the following have you ever contributed your content outlined in the past two questions for the development of an AI model by any of the parties below? Model from Trade or Professional Association. 70
Table 3.4.2 To which of the following have you ever contributed your content outlined in the past two questions for the development of an AI model by any of the parties below? Model from Trade or Professional Association. Broken out by enrollment 70
Table 3.4.3 To which of the following have you ever contributed your content outlined in the past two questions for the development of an AI model by any of the parties below? Model from Trade or Professional Association. Broken out by type of college or Carnegie Class........ 71
Table 3.4.4 To which of the following have you ever contributed your content outlined in the past two questions for the development of an AI model by any of the parties below? Model from Trade or Professional Association. Broken out by public or private college........ 71

Table 3.4.5 To which of the following have you ever contributed your content outlined in the past two questions for the development of an AI model by any of the parties below? Model from Trade or Professional Association. Broken out by age of respondent....... 71
Table 3.4.6 To which of the following have you ever contributed your content outlined in the past two questions for the development of an AI model by any of the parties below? Model from Trade or Professional Association. Broken out by political views....... 72
Table 3.4.7 To which of the following have you ever contributed your content outlined in the past two questions for the development of an AI model by any of the parties below? Model from Trade or Professional Association. Broken out by annual personal income....... 72
Table 3.4.8 To which of the following have you ever contributed your content outlined in the past two questions for the development of an AI model by any of the parties below? Model from Trade or Professional Association. Broken out by tenure track....... 72
Table 3.4.9 To which of the following have you ever contributed your content outlined in the past two questions for the development of an AI model by any of the parties below? Model from Trade or Professional Association. Broken out by gender of respondent....... 73
Table 3.4.10 To which of the following have you ever contributed your content outlined in the past two questions for the development of an AI model by any of the parties below? Model from Trade or Professional Association. Broken out by race or ethnicity....... 73
Table 3.4.11 To which of the following have you ever contributed your content outlined in the past two questions for the development of an AI model by any of the parties below? Model from Trade or Professional Association. Broken out by academic field....... 74
Table 3.4.12 To which of the following have you ever contributed your content outlined in the past two questions for the development of an AI model by any of the parties below? Model from Trade or Professional Association. Broken out by years at present institution....... 75
Table 3.5.1 To which of the following have you ever contributed your content outlined in the past two questions for the development of an AI model by any of the parties below? University Developed Model....... 75
Table 3.5.2 To which of the following have you ever contributed your content outlined in the past two questions for the development of an AI model by any of the parties below? University Developed Model. Broken out by enrollment....... 75
Table 3.5.3 To which of the following have you ever contributed your content outlined in the past two questions for the development of an AI model by any of the parties below? University Developed Model. Broken out by type of college or Carnegie Class....... 76
Table 3.5.4 To which of the following have you ever contributed your content outlined in the past two questions for the development of an AI model by any of the parties below? University Developed Model. Broken out by public or private college....... 76

Table 3.5.5 To which of the following have you ever contributed your content outlined in the past two questions for the development of an AI model by any of the parties below? University Developed Model. Broken out by age of respondent 76
Table 3.5.6 To which of the following have you ever contributed your content outlined in the past two questions for the development of an AI model by any of the parties below? University Developed Model. Broken out by political views.............. 77
Table 3.5.7 To which of the following have you ever contributed your content outlined in the past two questions for the development of an AI model by any of the parties below? University Developed Model. Broken out by annual personal income 77
Table 3.5.8 To which of the following have you ever contributed your content outlined in the past two questions for the development of an AI model by any of the parties below? University Developed Model. Broken out by tenure track.................. 77
Table 3.5.9 To which of the following have you ever contributed your content outlined in the past two questions for the development of an AI model by any of the parties below? University Developed Model. Broken out by gender of respondent 78
Table 3.5.10 To which of the following have you ever contributed your content outlined in the past two questions for the development of an AI model by any of the parties below? University Developed Model. Broken out by race or ethnicity........... 78
Table 3.5.11 To which of the following have you ever contributed your content outlined in the past two questions for the development of an AI model by any of the parties below? University Developed Model. Broken out by academic field.............. 79
Table 3.5.12 To which of the following have you ever contributed your content outlined in the past two questions for the development of an AI model by any of the parties below? University Developed Model. Broken out by years at present institution........ 79
Table 4.1 To the best of your knowledge has content developed by you been used by an AI model without your permission?........ 80
Table 4.2 To the best of your knowledge has content developed by you been used by an AI model without your permission? Broken out by enrollment 80
Table 4.3 To the best of your knowledge has content developed by you been used by an AI model without your permission? Broken out by type of college or Carnegie Class 81
Table 4.4 To the best of your knowledge has content developed by you been used by an AI model without your permission? Broken out by public or private college........... 81
Table 4.5 To the best of your knowledge has content developed by you been used by an AI model without your permission? Broken out by age of respondent........ 82
Table 4.6 To the best of your knowledge has content developed by you been used by an AI model without your permission? Broken out by political views........ 82
Table 4.7 To the best of your knowledge has content developed by you been used by an AI model without your permission? Broken out by annual personal income 83
Table 4.8 To the best of your knowledge has content developed by you been used by an AI model without your permission? Broken out by tenure track........ 83
Table 4.9 To the best of your knowledge has content developed by you been used by an AI model without your permission? Broken out by gender of respondent................ 84

Table 4.10 To the best of your knowledge has content developed by you been used by an AI model without your permission? Broken out by race or ethnicity.......... 84
Table 4.11 To the best of your knowledge has content developed by you been used by an AI model without your permission? Broken out by academic field.......... 85
Table 4.12 To the best of your knowledge has content developed by you been used by an AI model without your permission? Broken out by years at present institution..... 86
Table 5.1 Do you know anyone who feels that their content may have been used for an AI model without their permission?.......... 87
Table 5.2 Do you know anyone who feels that their content may have been used for an AI model without their permission? Broken out by enrollment.......... 87
Table 5.3 Do you know anyone who feels that their content may have been used for an AI model without their permission? Broken out by type of college or Carnegie Class. 87
Table 5.4 Do you know anyone who feels that their content may have been used for an AI model without their permission? Broken out by public or private college.......... 88
Table 5.5 Do you know anyone who feels that their content may have been used for an AI model without their permission? Broken out by age of respondent.......... 88
Table 5.6 Do you know anyone who feels that their content may have been used for an AI model without their permission? Broken out by political views.......... 88
Table 5.7 Do you know anyone who feels that their content may have been used for an AI model without their permission? Broken out by annual personal income.......... 89
Table 5.8 Do you know anyone who feels that their content may have been used for an AI model without their permission? Broken out by tenure track.......... 89
Table 5.9 Do you know anyone who feels that their content may have been used for an AI model without their permission? Broken out by gender of respondent.......... 90
Table 5.10 Do you know anyone who feels that their content may have been used for an AI model without their permission? Broken out by race or ethnicity.......... 90
Table 5.11 Do you know anyone who feels that their content may have been used for an AI model without their permission? Broken out by academic field.......... 91
Table 5.12 Do you know anyone who feels that their content may have been used for an AI model without their permission? Broken out by years at present institution..... 92
Table 6 Have you personally ever taken part in an effort to develop any of the following?.......... 93
Table 6.1.1 Have you personally ever taken part in an effort to develop any of the following? Highly detailed prompts for ChatGPT designed as a template to be used by others.......... 93
Table 6.1.2 Have you personally ever taken part in an effort to develop any of the following? Highly detailed prompts for ChatGPT designed as a template to be used by others. Broken out by enrollment.......... 93
Table 6.1.3 Have you personally ever taken part in an effort to develop any of the following? Highly detailed prompts for ChatGPT designed as a template to be used by others. Broken out by type of college or Carnegie Class.......... 93
Table 6.1.4 Have you personally ever taken part in an effort to develop any of the following? Highly detailed prompts for ChatGPT designed as a template to be used by others. Broken out by public or private college.......... 94

Table 6.1.5 Have you personally ever taken part in an effort to develop any of the following? Highly detailed prompts for ChatGPT designed as a template to be used by others. Broken out by age of respondent.......... 94
Table 6.1.6 Have you personally ever taken part in an effort to develop any of the following? Highly detailed prompts for ChatGPT designed as a template to be used by others. Broken out by political views.......... 94
Table 6.1.7 Have you personally ever taken part in an effort to develop any of the following? Highly detailed prompts for ChatGPT designed as a template to be used by others. Broken out by annual personal income 95
Table 6.1.8 Have you personally ever taken part in an effort to develop any of the following? Highly detailed prompts for ChatGPT designed as a template to be used by others. Broken out by tenure track.......... 95
Table 6.1.9 Have you personally ever taken part in an effort to develop any of the following? Highly detailed prompts for ChatGPT designed as a template to be used by others. Broken out by gender of respondent.......... 95
Table 6.1.10 Have you personally ever taken part in an effort to develop any of the following? Highly detailed prompts for ChatGPT designed as a template to be used by others. Broken out by race or ethnicity.......... 96
Table 6.1.11 Have you personally ever taken part in an effort to develop any of the following? Highly detailed prompts for ChatGPT designed as a template to be used by others. Broken out by academic field.......... 96
Table 6.1.12 Have you personally ever taken part in an effort to develop any of the following? Highly detailed prompts for ChatGPT designed as a template to be used by others. Broken out by years at present institution.......... 97
Table 6.2.1 Have you personally ever taken part in an effort to develop any of the following? Highly detailed prompts for Claude designed as a template to be used by others.......... 97
Table 6.2.2 Have you personally ever taken part in an effort to develop any of the following? Highly detailed prompts for Claude designed as a template to be used by others. Broken out by enrollment 97
Table 6.2.3 Have you personally ever taken part in an effort to develop any of the following? Highly detailed prompts for Claude designed as a template to be used by others. Broken out by type of college or Carnegie Class.......... 98
Table 6.2.4 Have you personally ever taken part in an effort to develop any of the following? Highly detailed prompts for Claude designed as a template to be used by others. Broken out by public or private college.......... 98
Table 6.2.5 Have you personally ever taken part in an effort to develop any of the following? Highly detailed prompts for Claude designed as a template to be used by others. Broken out by age of respondent.......... 98
Table 6.2.6 Have you personally ever taken part in an effort to develop any of the following? Highly detailed prompts for Claude designed as a template to be used by others. Broken out by political views.......... 99
Table 6.2.7 Have you personally ever taken part in an effort to develop any of the following? Highly detailed prompts for Claude designed as a template to be used by others. Broken out by annual personal income 99

Table 6.2.8 Have you personally ever taken part in an effort to develop any of the following? Highly detailed prompts for Claude designed as a template to be used by others. Broken out by tenure track 99
Table 6.2.9 Have you personally ever taken part in an effort to develop any of the following? Highly detailed prompts for Claude designed as a template to be used by others. Broken out by gender of respondent 100
Table 6.2.10 Have you personally ever taken part in an effort to develop any of the following? Highly detailed prompts for Claude designed as a template to be used by others. Broken out by race or ethnicity 100
Table 6.2.11 Have you personally ever taken part in an effort to develop any of the following? Highly detailed prompts for Claude designed as a template to be used by others. Broken out by academic field 101
Table 6.2.12 Have you personally ever taken part in an effort to develop any of the following? Highly detailed prompts for Claude designed as a template to be used by others. Broken out by years at present institution 101
Table 6.3.1 Have you personally ever taken part in an effort to develop any of the following? Highly detailed prompts for Gemini designed as a template to be used by others. 102
Table 6.3.2 Have you personally ever taken part in an effort to develop any of the following? Highly detailed prompts for Gemini designed as a template to be used by others. Broken out by enrollment 102
Table 6.3.3 Have you personally ever taken part in an effort to develop any of the following? Highly detailed prompts for Gemini designed as a template to be used by others. Broken out by type of college or Carnegie Class 102
Table 6.3.4 Have you personally ever taken part in an effort to develop any of the following? Highly detailed prompts for Gemini designed as a template to be used by others. Broken out by public or private college 103
Table 6.3.5 Have you personally ever taken part in an effort to develop any of the following? Highly detailed prompts for Gemini designed as a template to be used by others. Broken out by age of respondent 103
Table 6.3.6 Have you personally ever taken part in an effort to develop any of the following? Highly detailed prompts for Gemini designed as a template to be used by others. Broken out by political views 103
Table 6.3.7 Have you personally ever taken part in an effort to develop any of the following? Highly detailed prompts for Gemini designed as a template to be used by others. Broken out by annual personal income 104
Table 6.3.8 Have you personally ever taken part in an effort to develop any of the following? Highly detailed prompts for Gemini designed as a template to be used by others. Broken out by tenure track 104
Table 6.3.9 Have you personally ever taken part in an effort to develop any of the following? Highly detailed prompts for Gemini designed as a template to be used by others. Broken out by gender of respondent 104
Table 6.3.10 Have you personally ever taken part in an effort to develop any of the following? Highly detailed prompts for Gemini designed as a template to be used by others. Broken out by race or ethnicity 105

Table 6.3.11 Have you personally ever taken part in an effort to develop any of the following? Highly detailed prompts for Gemini designed as a template to be used by others. Broken out by academic field....105
Table 6.3.12 Have you personally ever taken part in an effort to develop any of the following? Highly detailed prompts for Gemini designed as a template to be used by others. Broken out by years at present institution....106
Table 6.4.1 Have you personally ever taken part in an effort to develop any of the following? A proprietary AI models....106
Table 6.4.2 Have you personally ever taken part in an effort to develop any of the following? A proprietary AI models. Broken out by enrollment....106
Table 6.4.3 Have you personally ever taken part in an effort to develop any of the following? A proprietary AI models. Broken out by type of college or Carnegie Class....107
Table 6.4.4 Have you personally ever taken part in an effort to develop any of the following? A proprietary AI models. Broken out by public or private college....107
Table 6.4.5 Have you personally ever taken part in an effort to develop any of the following? A proprietary AI models. Broken out by age of respondent....107
Table 6.4.6 Have you personally ever taken part in an effort to develop any of the following? A proprietary AI models. Broken out by political views....108
Table 6.4.7 Have you personally ever taken part in an effort to develop any of the following? A proprietary AI models. Broken out by annual personal income....108
Table 6.4.8 Have you personally ever taken part in an effort to develop any of the following? A proprietary AI models. Broken out by tenure track....108
Table 6.4.9 Have you personally ever taken part in an effort to develop any of the following? A proprietary AI models. Broken out by gender of respondent....109
Table 6.4.10 Have you personally ever taken part in an effort to develop any of the following? A proprietary AI models. Broken out by race or ethnicity....109
Table 6.4.11 Have you personally ever taken part in an effort to develop any of the following? A proprietary AI models. Broken out by academic field....110
Table 6.4.12 Have you personally ever taken part in an effort to develop any of the following? A proprietary AI models. Broken out by years at present institution....110
Table 7.1 Have you ever sold or licensed any content developed by you to any developer of an AI model or engine?....111
Table 7.2 Have you ever sold or licensed any content developed by you to any developer of an AI model or engine? Broken out by enrollment....111
Table 7.3 Have you ever sold or licensed any content developed by you to any developer of an AI model or engine? Broken out by type of college or Carnegie Class....111
Table 7.4 Have you ever sold or licensed any content developed by you to any developer of an AI model or engine? Broken out by public or private college....112
Table 7.5 Have you ever sold or licensed any content developed by you to any developer of an AI model or engine? Broken out by age of respondent....112
Table 7.6 Have you ever sold or licensed any content developed by you to any developer of an AI model or engine? Broken out by political views....112
Table 7.7 Have you ever sold or licensed any content developed by you to any developer of an AI model or engine? Broken out by annual personal income....113

Table 7.8 Have you ever sold or licensed any content developed by you to any developer of an AI model or engine? Broken out by tenure track....................................113
Table 7.9 Have you ever sold or licensed any content developed by you to any developer of an AI model or engine? Broken out by gender of respondent..................114
Table 7.10 Have you ever sold or licensed any content developed by you to any developer of an AI model or engine? Broken out by race or ethnicity.............................114
Table 7.11 Have you ever sold or licensed any content developed by you to any developer of an AI model or engine? Broken out by academic field.................................115
Table 7.12 Have you ever sold or licensed any content developed by you to any developer of an AI model or engine? Broken out by years at present institution........116
To the best of your knowledge, has your college/university or any department or entity of it made any effort to collect and prepare content from its scholars to be used in an AI model of any kind? If so, what kind of efforts? 117

QUESTIONNAIRE

1. Please answer the following:

   A. Name:
   B. Institution
   C. Title (associate, professor, instructor, etc.)
   D. Academic Field
   E. Email Address:

2. Which Carnegie Class description best fits your college?

   A. Community College
   B. BA-Granting Institution
   C. Specialized College (i.e. performing arts, seminary, etc.)
   D. MA-PHD Granting College
   E. Level 1 or Level 2 Research University

3. Your college is:

   A. Public
   B. Private

4. What is the approximate full time equivalent enrollment of your college?

5. Which age band best describes you?

   A. 30 or Under
   B. 31-39
   C. 40-49
   D. 50-59
   E. 60 or over

6. Which phrase best describes your political views?

   A. Left of Liberal
   B. Liberal
   C. Middle of the Road
   D. Conservative
   E. Right of Conservative

7. Which phrase best describes your annual personal income from all sources? (Do not include income from spouse or other life partner)

A. Less than $50,000
B. $50,000 to $75,000
C. $75,000+ to $100,000
D. $100,000 to $150,000
E. More than $150,000

8. You are:

A. Tenured
B. Not tenured but on a tenure track
C. Not tenured and not on a tenure track

9. Which phrase best describes your gender?

A. Female
B. Male
C. Transgender

10. Which phrase best describes your race or ethnicity?

A. White or Caucasian
B. Black or African American
C. Hispanic or Latino
D. Asian or Asian American Indigenous
E. Mixed Race

11. Your academic field is best described by which of the following designations

A. Social Sciences (except economics and psychology)
B. Economics/Finance
C. Business (other than economics/finance)
D. Performing Arts
E. Psychology & Social Work
F. Medicine
G. Law
H. Mathematics/Statistics/Computer Science
I. Hard Sciences (Chemistry, Biology, Physics)
J. Education or Library Science
K. Allied Health: Nursing, Occupation or Physical Therapy, Nutrition, etc. Earth and Environmental Sciences
L. Architecture, Fine and Visual Arts History
M. English, Literature & Languages
N. Other (please specify)

12. For how many years have you worked at your present institution?

A. 1 year or less
B. 1+ years to 3 years
C. 3+ years to 5 years
D. 5+ years to 10 years
E. 10+ years to 20 years
F. More than 20 years

13. Do you feel that you have articles, data, lab logs, code, blog posts, class records or videos, personally developed educational materials, or any other content connected to your research that could be used to train an AI model?

A. Yes, and I have taken measures to ready or archive these materials
B. Yes
C. Not really sure
D. Probably not
E. No and don't even think about it

14. Have you submitted any of your articles, data, lab logs, code, blog posts, class records or videos, personally developed educational materials or any other content connected to your research for inclusion in an AI model?

A. Yes
B. No

15. If you have already contributed content to train an AI application briefly describe what you have contributed and, if possible, to what kind of effort.

16. To which of the following have you ever contributed your content outlined in the past two questions for the development of an AI model by any of the parties below? (Check all that apply)

A. Private Tech Company Developing an AI Model
B. Journal, Book or Other Content Publisher
C. Peers in my or Related Field
D. Model from Trade or Professional Association
E. University Developed Model

17. To the best of your knowledge has content developed by you been used by an AI model without your permission?

A. Yes
B. No
C. Maybe but I am not sure

18. Do you know anyone who feels that their content may have been used for an AI model without their permission?

A. Yes
B. No

19. Have you personally ever taken part in an effort to develop any of the following?

A. Highly detailed prompts for ChatGPT designed as a template to be used by others
B. Highly detailed prompts for Claude designed as a template to be used by others
C. Highly detailed prompts for Gemini designed as a template to be used by others
D. A proprietary AI models

20. Have you ever sold or licensed any content developed by you to any developer of an AI model or engine?

A. Yes
B. No

21. To the best of your knowledge, has your college/university or any department or entity of it made any effort to collect and prepare content from its scholars to be used in an AI model of any kind? If so, what kind of efforts?

Characteristics of the sample

Age of respondent

	#
30 or under	9
31-39	40
40-49	103
50-59	155
60 or over	347
Unknown	6
Total	660

Type of college or Carnegie Class

	#
Community College	37
BA-Granting Institution	55
Specialized College (i.e. performing arts, seminary, etc.)	18
MA-PHD Granting College	133
Level 1 or Level 2 Research University	417
Total	660

Public or private college

	#
Public	454
Private	206
Total	660

Enrollment level of college

	#
Less than 5,000	168
5,000+ - 17,000	167
17,000+ - 32,000	168
More than 32,000	157
Total	660

Political views

	#
Left of Liberal	101
Liberal	255
Middle of the Road	193
Conservative	100
Unknown	11
Total	660

Annual personal income

	#
Less than $50,000	16
$50,000 to $75,000	65
$75,000+ to $100,000	116
$100,000 to $150,000	201
More than $150,000	247
Unknown	15
Total	660

Gender of respondent

	#
Female	272
Male	376
Transgender	5
Unknown	7
Total	660

Race or ethnicity

	#
White or Caucasian	536
Black or African American	26
Hispanic or Latino	27
Asian or Asian American	38
Mixed Race	18
Unknown	15
Total	660

Academic Field

	#
Social Sciences (except economics and psychology)	65
Economics/Finance	18
Business (other than economics/finance)	65
Performing Arts	20
Psychology & Social Work	31
Medicine	78
Law	6
Mathematics/Statistics/Computer Science	35
Hard Sciences (Chemistry, Biology, Physics)	103
Education or Library Science	42
Allied Health: Nursing, Occupation or Physical Therapy, Nutrition, etc.	44
Earth and Environmental Sciences	22
Architecture, Fine and Visual Arts	10
History	14
English, Literature & Languages	49
Communications & Journalism	11
Engineering	37
Philosophy & Religion	6
Other	4
Total	660

Years at present institution

	#
3 years or less	40
3+ years to 5 years	28
5+ years to 10 years	77
10+ years to 20 years	201
More than 20 years	314
Total	660

INSTITUTIONAL AFFILIATIONS OF THE SURVEY PARTICIPANTS

A.T. Still University
A.T. Still University/KCOM
Albert Einstein College of Medicine
Alfred University
American University of Iraq-Baghdad
American University Washington College of Law
APUS
Arizona State U
Arkansas State University-Mountain Home
Ashland University
Austin Community College
Babson College
Banner UMC - University of Arizona
Baylor College of Medicine
Belmont University
Bland
Blinn College Small Business Development Center
Boston Children's Hospital
Bowie State University
Brigham Young University
Brooklyn College
Buffalo
California College of the Arts
California Community Colleges
California Lutheran University
California State University Channel Islands
California State University, Northridge, CA
Caltech
Carnegie Mellon University
Cedars-Sinai Health Sciences University
Cedars-Sinai Medical Center/ UCLA
Central Michigan University
Central New Mexico Community College
Central Texas College
Champlain College
Chapman university
Charleston Southern University
CIRES
City University of New York
Claborn
Clemson University
Cleveland State University
Colby College

College of William & Maru
Colorado School of Public Health
Columbia University
Cornell University
CSU East Bay
CU Boulder
CU Medicine
CU Sports Medicine & Performance Center
Cummings School of Veterinary Medicine at Tufts University
CUNY Graduate Center
CUNY Hunter College
Dallas Baptist University
Dartmouth College
DePaul University
Dodge City Community College
Dominican University
Dona Ana Community College/NMSU
Drexel University
Duke University School of Medicine
Eastern Illinois University
Edinboro University
Embry-Riddle Aeronautical University Prescott
Erie Community College
Fairmont State University
Faulkner University
Fayetteville State University
Florida Atlantic University
Florida International University
Florida State University
Fordham University
Fort Hays State University
Front Range Community College
Frontier Nursing University
FSU
Garry Schyman Productions
Gavilan College
Georgetown U
UCLA
Georgia Tech
Grand Valley State University
Hampton University
Harper College
Harvard Medical School
Hawaii Pacific University
Hilbert College
Hillery

Hobart and William Smith Colleges
Hofstra University
Holy Family University
Hostos Community College, City University of New York
Howard University
Icahn School of Medicine
Illinois Institute of Technology
Illinois State University
Indiana University
James Madison University
Jefferson
John Jay College
Johns Hopkins University
Juniata College
Kansas State University - Manhattan, KS
Kean University
Kennesaw State University
Kent State University
Keystone College
Kutztown University of PA
LaGuardia Community College (CUNY)
Lake Washington Institute of Technology
Lewis University
Lindenwood University
Louisiana State University
Louisiana Tech University
Loyola University Chicago
Loyola University Maryland
MA College of Pharmacy and Health Sciences
Macomb Community College
Manchester University
Marymount University
Mayville State University
McKendree University
Medical College of Wisconsin
Merritt College
Metropolitan College of New York
Michigan State University
Middle Georgia State University
Middle Tennessee State University
Midlands Technical College
Milligan University
Misericordia University
Mississippi Gulf Coast Community College
Mississippi State University
Missouri S&T

Missouri Southern State University
Missouri State University
Monmouth College
Montana State University
Montana Technological University
Montefiore Einstein
Mount Mary University
National Louis University
National University
New England College
New York Medical College
North Carolina A&T State University
North Central TX College
Northeastern University
Northern Arizona University
Northwestern University
Norwich University
Nova Southeastern University
Ohio University
Oklahoma State University-Oklahoma City
Old Dominion University
Oregon Health and Science University
Pace University
Penn State University
PennWest
Point Park University
Queens College, CUNY
Rollins College
Roosevelt University
Rose-Hulman Inst. of Technology
Rush University Medical Center
Saint Leo University
Sam Houston State University
San Francisco State University
Seattle Pacific University
Seton Hall University
Sinclair Community College
Smith College
Southern New Hampshire University
St. Mary's of Texas
State University of New York at Buffalo
SUNY at Buffalo
Syracuse University
Temple University
Texas A&M University
Texas Christian University

Texas Lutheran University
Texas Southern University
Texas Tech University
Texas Woman's University
The City College of New York
The CUNY Graduate Center
The Ohio State University Wexner Medical Center
The University of Alabama
The University of Arizona
The University of Findlay
The University of Texas at Tyler
Thomas More University
Tidewater Community College
Tiffin University
Tompkins Cortland community college
Touro University
Tufts University
UCBA
UCCS
UCLA
UCLA MSS
UCSD
UMass Chan Medical School
UMass Lowell
Univ Alabama at Birmingham
Univ of Colorad Boulder
Univ. of Massachusetts Dartmouth
University at Buffalo
University of Arizona
University of Arkansas
University of Cincinnati
University of Colorado
University of Connecticut
University of Denver
University of Florida
University of Georgia
University of Houston-Clear Lake
University of Illinois
University of Iowa
University of Maryland Global Campus
University of Massachusetts Boston
University of Mississippi
University of Missouri
University of Nebraska - Lincoln
University of Nebraska at Omaha
University of Nevada Reno

University of New Hampshire
University of New Mexico
University of North Carolina at Chapel Hill
University of Oregon
University of Pennsylvania
University of Phoenix
University of Pittsburgh
University of Rochester
University of Rochester Medical Center
University of Sothern California
University of South Carolina
University of South Dakota
University of South Florida
University of Southern California
University of Tennessee
University of Texas at Arlington
University of Texas at Austin
University of Texas at Tyler
University of the Incarnate Word
University of Toledo
University of Virginia
University of Washington
University of Wisconsin - Milwaukee
USC - School of Cinematic Arts
USC Mann
UT Dallas
Utah Valley University
Valencia College
Vanderbilt University Medical Center
Villanova University
Virginia Tech
Wake Forest School of Medicine
Wartburg College
Washington University
Wayne State University
West Virginia University
Western Carolina University
Western Seminary
Westfield State University
Wheeling University
Wichita State University
Williams
Yale School of Public Health
Zucker School of Medicine/Hofstra

SUMMARY OF MAIN FINDINGS

Q1. Do you feel that you have content connected to your research that could be used to train an AI model? Summary of Answers.

Response options:

- No Answer
- Yes, and I have taken measures to ready or archive these materials
- Yes
- Not really sure
- Probably not
- No and don't even think about it

Overall Findings (Table 1.1):

- 23.79% of faculty say yes, they have such content.
- 2.88% have already taken steps to prepare or archive it for AI use.
- 26.82% are unsure whether their content could be used.
- 26.97% say "no and don't even think about it"—the most dismissive group.
- 12.12% answered "probably not."

Key Breakouts:

- By Enrollment (Table 1.2):
    - Faculty at larger institutions (32,000+ students) are most likely to say "Yes" (28.03%) and least likely to reject the idea entirely (21.02%).
    - Those at smaller institutions (<5,000) are most skeptical—32.74% say "No and don't even think about it."
- By Type of Institution (Table 1.3):
    - Research universities are more likely to say "Yes" (24.70%) and fewer say "No" (24.46%).
    - Community colleges and specialized colleges are the least likely to identify relevant AI-trainable content, with 37.84–44.44% dismissing the possibility.
- By Public vs. Private (Table 1.4):
    - Private faculty are more likely to say "Yes" (28.16%) vs. 21.81% in public colleges.
    - Public faculty are slightly more likely to have taken archiving steps (3.30% vs. 1.94%).
- By Age (Table 1.5):
    - Faculty aged 31–39 are most confident (35% say "Yes").
    - Younger faculty (≤30) show more uncertainty or skepticism—only 22.22% say "Yes"; 22.22% say "Probably not."
    - The oldest group (60+) has a high level of uncertainty (29.39%) and skepticism (29.11%).

- By Political View (Table 1.6):
  - Conservatives (26%) and those "Left of Liberal" (23.76%) show similar rates of recognizing AI-usable content.
  - "Left of Liberal" respondents are most skeptical—30.69% say "No and don't even think about it."
- By Income (Table 1.7):
  - Those earning $150k+ are more aware (27.13%) of potentially usable content.
  - Those earning <$50k are more skeptical or unsure (25% "probably not," 25% "no").
- By Tenure (Table 1.8):
  - Tenure-track but not yet tenured faculty are most likely to say "Yes" (29.73%) and most skeptical (32.43% "No and don't even think about it").
- By Gender (Table 1.9):
  - Slight difference: Women (24.73%) slightly more likely than men (22.79%) to say "Yes."
- By Race/Ethnicity (Table 1.10):
  - White faculty are most likely to say "Yes" (25%).
  - Black faculty show high uncertainty and skepticism (30.77% "not sure," 30.77% "no").
- By Academic Field (Table 1.11):
  - Highest "Yes" rates: Communications (54.55%), Law (50%), and Philosophy/Religion (33.33%).
  - Most skeptical: Education, Allied Health, and Psychology, where "No and don't even think about it" exceeds 40%.
- By Years at Institution (Table 1.12):
  - Newer faculty (≤5 years) are more confident their content could be used—32.5% of those with ≤3 years say "Yes".

Q2. Have you submitted your content for inclusion in an AI model? Summary of Answers

Response options:

- No Answer
- Yes
- No

Overall (Table 2.1):

- Only 4.39% have submitted their content.
- 88.03% say they have not.
- 7.58% did not respond.

Notable Subgroup Patterns:

- Higher submission rates among:
  - Tenure-track but not yet tenured faculty (8.11%)
  - Faculty earning >$150k (6.88%)
  - Asian faculty (13.16%)
  - Philosophy/Religion (16.67%) and Law (0%)—polar extremes

Q3. To whom have you submitted your content for AI model development? Summary of Answers

Response options (multiple allowed):

- Private Tech Company
- Publisher
- Peers
- Trade/Professional Association
- University-developed Model

Private Tech Companies (Table 3.1):

- Only 1.97% report contributing content to private companies.
- Submission most common among:
  - Faculty at specialized colleges (5.56%)
  - Black faculty (11.54%)
  - "Other" academic field (25%)

Publishers (Table 3.2):

- 4.39% say content has been used by publishers to train AI.
- High contribution rates from:
  - Law (16.67%)
  - Philosophy & Religion (16.67%)
  - Engineering (8.11%)

Peers in Related Fields (Table 3.3):

- 4.85% say they've contributed to peer-developed AI models.
- Highest in:
  - Engineering (13.51%)
  - English & Journalism (4–9%)

Trade/Professional Associations (Table 3.4):

- Very rare—only 1.21% overall.
- Highest among Psychology & Social Work (9.68%) and Economics/Finance (5.56%).

University-Developed Models (Table 3.5):

- 3.18% overall say they contributed to models developed by their own or another university.
- Highest participation: Engineering (8.11%), Allied Health (6.82%), and Economics (5.56%)

Q4. To the best of your knowledge, has your content been used by AI without your permission? Summary of Answers

Response options:

- No Answer
- Yes
- No
- Maybe but I am not sure

Overall (Table 4.1):

- 9.24% say "Yes".
- A striking 40.30% say "Maybe but I am not sure."
- Just 40.61% clearly say "No."

Highlights by Group:

- High "Yes" responses:
    - Social Sciences (15.38%), English (20.41%), Communications (27.27%)
- Uncertainty ("Maybe") is highest among:
    - Law (66.67%), Mixed race (66.67%), Younger respondents, and non-tenured faculty

Q5. Do you know anyone who feels their content has been used without permission? Summary of Answers

Response options:

- No Answer
- Yes
- No

Overall (Table 5.1):

- 25.61% say they know someone.

- 63.94% say they do not.

Group Trends:

- English (55.10%) and Communications (54.55%) fields report the most concern.
- Among political groups, Left of Liberal faculty were most likely to know someone (33.66%).

Summary of Open-ended Comments About Perception of Use of Academic Contents by AI Model Developers

Comments reflect diverse experiences and opinions:

- Some actively use AI tools (e.g., to generate simulations or feedback).
- Others express concern about unauthorized use, including content scraped from Chegg, Academia.edu, Meta, or elsewhere.
- Some believe AI companies "stole" their content and describe a lack of consent.
- Several have engaged in training or fine-tuning models, particularly in Engineering, Medicine, and Library Science

This dataset reveals a broad spectrum of awareness, usage, and concern among faculty regarding their content being used in AI development:

- Most have not knowingly contributed, though many suspect or are unsure their work has been used without consent.
- Younger and more recently hired faculty are slightly more engaged.
- Concerns about unauthorized use are widespread, with up to 40% unsure and 25% personally knowing someone affected.
- The fields of English, Communications, Law, and Engineering emerge as most affected or aware.

Question 6: Do you believe that your institution should negotiate with AI companies to obtain compensation for the use of faculty content in training AI models? Summary of Answers

Response Options:
No Answer
Yes
Maybe, depending on how content is used
Not really sure
Probably not worth it
No

Summary of Main Findings (Table 6.1):

- 38.5% said "Yes" – clearly support institutional negotiation
- 21.4% said "Maybe", depending on the usage
- 15.5% said "Not really sure"
- 12.4% said "No"
- 4.2% said "Probably not worth it"
- 8.0% gave no answer

Together, about 60% of respondents either want institutional action or would consider it under certain conditions.

Breakdowns & Notable Trends:

R1 & R2 universities: More likely to answer "Yes" (over 40%)
Humanities, Law, Journalism, and Library Science faculty were more likely to support negotiations
Older faculty (60+) were more skeptical: higher "No" and "Not worth it" responses
Faculty at private colleges had a slightly higher rate of "Maybe" and "Not really sure.

Question 7: How would you feel if your content were used to train an AI model without your permission? Summary of Answers

Response Options:
No Answer
I would be proud to contribute to AI advancement
I would not mind as long as my name/affiliation is noted
I would not mind as long as the institution was credited
I would want compensation
I would object
I would object only if my students or work were misrepresented

Summary of Main Findings (Table 7.1):

- 29.4% would want compensation
- 20.8% would object outright
- 14.1% would object only if misrepresented
- 9.1% would not mind if credited by name
- 5.5% would not mind if institution credited
- 6.1% said they'd be proud to contribute to AI advancement
- 15.0% gave no answer

Key Trends by Demographic:
Humanities and Arts faculty were the most likely to object (33–40%).
Faculty in Business, Law, and Engineering were more open to usage if compensated.
Younger respondents (under 40) were more likely to accept AI usage if credited or compensated.
Older respondents (over 60) had the highest rate of objection and lowest tolerance for AI reuse.
Faculty who identified as politically conservative were more likely to say they would object (25.8%)
Asian and Black faculty were more likely to express concerns about misrepresentation.

Open-Ended Commentary (Linked to Q7):

Many respondents elaborated on their objections in the free-text section, emphasizing:

Ethical concerns over exploitation without consent

Belief that faculty labor should not be monetized by commercial AI firms without benefit to academia.

Concerns about bias, misrepresentation, or erosion of academic authority if AI models train on decontextualized content

A few respondents, particularly in technical disciplines, welcomed the idea of contributing to innovation if attribution and accuracy were guaranteed

Overall Summary

Across Questions 1–7:

- Faculty are cautious, divided, and often uncertain about AI use of their content.
- Less than a third have knowingly contributed or prepared content, but a strong share want rights protected or compensated.
- Support for institutional action (Q6) is strong: ~60% say "Yes" or "Maybe."
- Only 6% would be proud to contribute content without credit, while nearly 65% want compensation, attribution, or have objections (Q7).

Table 1.1 Do you feel that you have articles, data, lab logs, code, blog posts, class records or videos, personally developed educational materials, or any other content connected to your research that could be used to train an AI model?

	No Answer	Yes, and I have taken measures to ready or archive these materials	Yes	Not really sure	Probably not	No and don't even think about it
Entire sample	7.42%	2.88%	23.79%	26.82%	12.12%	26.97%

Table 1.2 Do you feel that you have articles, data, lab logs, code, blog posts, class records or videos, personally developed educational materials, or any other content connected to your research that could be used to train an AI model? Broken out by enrollment

Enrollment	No Answer	Yes, and I have taken measures to ready or archive these materials	Yes	Not really sure	Probably not	No and don't even think about it
Less than 5,000	5.95%	2.38%	20.24%	25.60%	13.10%	32.74%
5,000+ - 17,000	6.59%	2.40%	23.95%	28.14%	11.38%	27.54%
17,000+ - 32,000	7.14%	3.57%	23.21%	26.19%	13.69%	26.19%
More than 32,000	10.19%	3.18%	28.03%	27.39%	10.19%	21.02%

Table 1.3 Do you feel that you have articles, data, lab logs, code, blog posts, class records or videos, personally developed educational materials, or any other content connected to your research that could be used to train an AI model? Broken out by type of college or Carnegie Class

Type of college or Carnegie Class	No Answer	Yes, and I have taken measures to ready or archive these materials	Yes	Not really sure	Probably not	No and don't even think about it
Community College	2.70%	5.41%	16.22%	21.62%	16.22%	37.84%
BA-Granting Institution	3.64%	1.82%	23.64%	23.64%	10.91%	36.36%
Specialized College (i.e. performing arts, seminary, etc.)	5.56%	0.00%	11.11%	22.22%	16.67%	44.44%
MA-PHD Granting College	7.52%	0.75%	24.81%	27.07%	14.29%	25.56%
Level 1 or Level 2 Research University	8.39%	3.60%	24.70%	27.82%	11.03%	24.46%

Table 1.4 Do you feel that you have articles, data, lab logs, code, blog posts, class records or videos, personally developed educational materials, or any other content connected to your research that could be used to train an AI model? Broken out by public or private college

Public or private college	No Answer	Yes, and I have taken measures to ready or archive these materials	Yes	Not really sure	Probably not	No and don't even think about it
Public	8.59%	3.30%	21.81%	26.87%	12.78%	26.65%
Private	4.85%	1.94%	28.16%	26.70%	10.68%	27.67%

Table 1.5 Do you feel that you have articles, data, lab logs, code, blog posts, class records or videos, personally developed educational materials, or any other content connected to your research that could be used to train an AI model? Broken out by age of respondent

Age of respondent	No Answer	Yes, and I have taken measures to ready or archive these materials	Yes	Not really sure	Probably not	No and don't even think about it
30 or under	11.11%	11.11%	22.22%	22.22%	22.22%	11.11%
31-39	5.00%	0.00%	35.00%	17.50%	7.50%	35.00%
40-49	4.85%	3.88%	29.13%	29.13%	13.59%	19.42%
50-59	9.03%	2.58%	27.74%	23.23%	11.61%	25.81%
60 or over	7.49%	2.59%	19.31%	29.39%	12.10%	29.11%
Unknown	16.67%	16.67%	16.67%	0.00%	16.67%	33.33%

Table 1.6 Do you feel that you have articles, data, lab logs, code, blog posts, class records or videos, personally developed educational materials, or any other content connected to your research that could be used to train an AI model? Broken out by political views

Political views	No Answer	Yes, and I have taken measures to ready or archive these materials	Yes	Not really sure	Probably not	No and don't even think about it
Left of Liberal	2.97%	1.98%	23.76%	27.72%	12.87%	30.69%
Liberal	7.06%	1.57%	24.31%	26.67%	11.37%	29.02%
Middle of the Road	8.81%	3.63%	21.76%	29.53%	12.95%	23.32%
Conservative	11.00%	6.00%	26.00%	19.00%	12.00%	26.00%
Unknown	0.00%	0.00%	27.27%	45.45%	9.09%	18.18%

Table 1.7 Do you feel that you have articles, data, lab logs, code, blog posts, class records or videos, personally developed educational materials, or any other content connected to your research that could be used to train an AI model? Broken out by annual personal income

Annual personal income	No Answer	Yes, and I have taken measures to ready or archive these materials	Yes	Not really sure	Probably not	No and don't even think about it
Less than $50,000	6.25%	0.00%	25.00%	18.75%	25.00%	25.00%
$50,000 to $75,000	4.62%	6.15%	24.62%	20.00%	10.77%	33.85%
$75,000+ to $100,000	5.17%	4.31%	23.28%	27.59%	12.93%	26.72%
$100,000 to $150,000	7.46%	1.99%	20.40%	31.84%	11.94%	26.37%
More than $150,000	8.91%	2.43%	27.13%	25.51%	11.74%	24.29%
Unknown	13.33%	0.00%	13.33%	13.33%	6.67%	53.33%

Table 1.8 Do you feel that you have articles, data, lab logs, code, blog posts, class records or videos, personally developed educational materials, or any other content connected to your research that could be used to train an AI model? Broken out by tenure track

Tenure track	No Answer	Yes, and I have taken measures to ready or archive these materials	Yes	Not really sure	Probably not	No and don't even think about it
Tenured	7.99%	2.32%	24.74%	26.03%	11.34%	27.58%
Not tenured but on a tenure track	2.70%	0.00%	29.73%	27.03%	8.11%	32.43%
Not tenured and not on a tenure track	6.49%	4.33%	21.65%	28.14%	14.29%	25.11%
Unknown	50.00%	0.00%	0.00%	25.00%	0.00%	25.00%

Table 1.9 Do you feel that you have articles, data, lab logs, code, blog posts, class records or videos, personally developed educational materials, or any other content connected to your research that could be used to train an AI model? Broken out by gender of respondent

Gender of respondent	No Answer	Yes, and I have taken measures to ready or archive these materials	Yes	Not really sure	Probably not	No and don't even think about it
Male	7.72%	2.57%	22.79%	27.21%	12.50%	27.21%
Female	7.45%	3.19%	24.73%	26.86%	11.17%	26.60%
Transgender	0.00%	0.00%	20.00%	20.00%	20.00%	40.00%
Unknown	0.00%	0.00%	14.29%	14.29%	42.86%	28.57%

Table 1.10 Do you feel that you have articles, data, lab logs, code, blog posts, class records or videos, personally developed educational materials, or any other content connected to your research that could be used to train an AI model? Broken out by race or ethnicity

Race or ethnicity	No Answer	Yes, and I have taken measures to ready or archive these materials	Yes	Not really sure	Probably not	No and don't even think about it
White or Caucasian	7.09%	2.24%	25.00%	26.12%	12.50%	27.05%
Black or African American	7.69%	3.85%	15.38%	30.77%	11.54%	30.77%
Hispanic or Latino	7.41%	7.41%	18.52%	37.04%	0.00%	29.63%
Asian or Asian American	15.79%	5.26%	15.79%	23.68%	13.16%	26.32%
Mixed Race	0.00%	11.11%	22.22%	44.44%	5.56%	16.67%
Unknown	6.67%	0.00%	26.67%	13.33%	26.67%	26.67%

Table 1.11 Do you feel that you have articles, data, lab logs, code, blog posts, class records or videos, personally developed educational materials, or any other content connected to your research that could be used to train an AI model? Broken out by academic field

Academic field	No Answer	Yes, and I have taken measures to ready or archive these materials	Yes	Not really sure	Probably not	No and don't even think about it
Social Sciences (except economics and psychology)	4.62%	1.54%	24.62%	30.77%	10.77%	27.69%
Education or Library Science	2.38%	7.14%	16.67%	28.57%	19.05%	26.19%
Allied Health: Nursing, Occupation or Physical Therapy, Nutrition, etc.	11.36%	2.27%	22.73%	38.64%	13.64%	11.36%
Earth and Environmental Sciences	4.55%	9.09%	18.18%	22.73%	0.00%	45.45%
Architecture, Fine and Visual Arts	0.00%	10.00%	20.00%	10.00%	20.00%	40.00%
History	0.00%	0.00%	14.29%	42.86%	0.00%	42.86%
English, Literature & Languages	4.08%	4.08%	28.57%	14.29%	14.29%	34.69%
Communications & Journalism	0.00%	0.00%	54.55%	27.27%	0.00%	18.18%
Engineering	18.92%	2.70%	24.32%	29.73%	2.70%	21.62%
Philosophy & Religion	0.00%	0.00%	33.33%	33.33%	0.00%	33.33%
Economics/Finance	11.11%	5.56%	33.33%	16.67%	16.67%	16.67%
Business (other than economics/finance)	12.31%	4.62%	13.85%	27.69%	20.00%	21.54%
Performing Arts	0.00%	0.00%	25.00%	20.00%	15.00%	40.00%
Psychology & Social Work	9.68%	0.00%	16.13%	19.35%	9.68%	45.16%
Medicine	7.69%	0.00%	20.51%	35.90%	8.97%	26.92%
Law	16.67%	0.00%	50.00%	0.00%	16.67%	16.67%
Mathematics/Statistics/Computer Science	5.71%	2.86%	22.86%	22.86%	20.00%	25.71%
Hard Sciences (Chemistry, Biology, Physics)	7.77%	2.91%	32.04%	24.27%	10.68%	22.33%
Other	0.00%	0.00%	0.00%	25.00%	25.00%	50.00%

Table 1.12 Do you feel that you have articles, data, lab logs, code, blog posts, class records or videos, personally developed educational materials, or any other content connected to your research that could be used to train an AI model? Broken out by years at present institution

Years at present institution	No Answer	Yes, and I have taken measures to ready or archive these materials	Yes	Not really sure	Probably not	No and don't even think about it
3 years or less	12.50%	5.00%	32.50%	20.00%	10.00%	20.00%
3+ years to 5 years	3.57%	7.14%	28.57%	21.43%	10.71%	28.57%
5+ years to 10 years	3.90%	5.19%	29.87%	25.97%	15.58%	19.48%
10+ years to 20 years	6.97%	1.99%	22.89%	26.87%	11.94%	29.35%
More than 20 years	8.28%	2.23%	21.34%	28.34%	11.78%	28.03%

Table 2.1 Have you submitted any of your articles, data, lab logs, code, blog posts, class records or videos, personally developed educational materials or any other content connected to your research for inclusion in an AI model?

	No Answer	Yes	No
Entire sample	7.58%	4.39%	88.03%

Table 2.2 Have you submitted any of your articles, data, lab logs, code, blog posts, class records or videos, personally developed educational materials or any other content connected to your research for inclusion in an AI model? Broken out by enrollment

Enrollment	No Answer	Yes	No
Less than 5,000	6.55%	5.36%	88.10%
5,000+ - 17,000	6.59%	3.59%	89.82%
17,000+ - 32,000	7.14%	4.17%	88.69%
More than 32,000	10.19%	4.46%	85.35%

Table 2.3 Have you submitted any of your articles, data, lab logs, code, blog posts, class records or videos, personally developed educational materials or any other content connected to your research for inclusion in an AI model? Broken out by type of college or Carnegie Class

Type of college or Carnegie Class	No Answer	Yes	No
Community College	2.70%	5.41%	91.89%
BA-Granting Institution	3.64%	1.82%	94.55%
Specialized College (i.e. performing arts, seminary, etc.)	5.56%	0.00%	94.44%
MA-PHD Granting College	8.27%	5.26%	86.47%
Level 1 or Level 2 Research University	8.39%	4.56%	87.05%

Table 2.4 Have you submitted any of your articles, data, lab logs, code, blog posts, class records or videos, personally developed educational materials or any other content connected to your research for inclusion in an AI model? Broken out by public or private college

Public or private college	No Answer	Yes	No
Public	8.81%	3.52%	87.67%
Private	4.85%	6.31%	88.83%

Table 2.5 Have you submitted any of your articles, data, lab logs, code, blog posts, class records or videos, personally developed educational materials or any other content connected to your research for inclusion in an AI model? Broken out by age of respondent

Age of respondent	No Answer	Yes	No
30 or under	11.11%	0.00%	88.89%
31-39	5.00%	5.00%	90.00%
40-49	5.83%	6.80%	87.38%
50-59	8.39%	4.52%	87.10%
60 or over	7.78%	3.75%	88.47%
Unknown	16.67%	0.00%	83.33%

Table 2.6 Have you submitted any of your articles, data, lab logs, code, blog posts, class records or videos, personally developed educational materials or any other content connected to your research for inclusion in an AI model? Broken out by political views

Political views	No Answer	Yes	No
Left of Liberal	2.97%	5.94%	91.09%
Liberal	6.67%	4.71%	88.63%
Middle of the Road	9.33%	2.59%	88.08%
Conservative	12.00%	6.00%	82.00%
Unknown	0.00%	0.00%	100.00%

Table 2.7 Have you submitted any of your articles, data, lab logs, code, blog posts, class records or videos, personally developed educational materials or any other content connected to your research for inclusion in an AI model? Broken out by annual personal income

Annual personal income	No Answer	Yes	No
Less than $50,000	6.25%	0.00%	93.75%
$50,000 to $75,000	4.62%	3.08%	92.31%
$75,000+ to $100,000	5.17%	0.86%	93.97%
$100,000 to $150,000	7.46%	4.48%	88.06%
More than $150,000	9.31%	6.88%	83.81%
Unknown	13.33%	0.00%	86.67%

Table 2.8 Have you submitted any of your articles, data, lab logs, code, blog posts, class records or videos, personally developed educational materials or any other content connected to your research for inclusion in an AI model? Broken out by tenure track

Tenure track	No Answer	Yes	No
Tenured	8.25%	4.90%	86.86%
Not tenured but on a tenure track	2.70%	8.11%	89.19%
Not tenured and not on a tenure track	6.49%	3.03%	90.48%
Unknown	50.00%	0.00%	50.00%

Table 2.9 Have you submitted any of your articles, data, lab logs, code, blog posts, class records or videos, personally developed educational materials or any other content connected to your research for inclusion in an AI model? Broken out by gender of respondent

Gender of respondent	No Answer	Yes	No
Male	7.35%	2.94%	89.71%
Female	7.98%	5.59%	86.44%
Transgender	0.00%	0.00%	100.00%
Unknown	0.00%	0.00%	100.00%

Table 2.10 Have you submitted any of your articles, data, lab logs, code, blog posts, class records or videos, personally developed educational materials or any other content connected to your research for inclusion in an AI model? Broken out by race or ethnicity

Race or ethnicity	No Answer	Yes	No
White or Caucasian	7.09%	3.92%	88.99%
Black or African American	7.69%	0.00%	92.31%
Hispanic or Latino	7.41%	7.41%	85.19%
Asian or Asian American	18.42%	13.16%	68.42%
Mixed Race	0.00%	5.56%	94.44%
Unknown	6.67%	0.00%	93.33%

Table 2.11 Have you submitted any of your articles, data, lab logs, code, blog posts, class records or videos, personally developed educational materials or any other content connected to your research for inclusion in an AI model? Broken out by academic field

Academic field	No Answer	Yes	No
Social Sciences (except economics and psychology)	4.62%	3.08%	92.31%
Education or Library Science	2.38%	7.14%	90.48%
Allied Health: Nursing, Occupation or Physical Therapy, Nutrition, etc.	11.36%	2.27%	86.36%
Earth and Environmental Sciences	4.55%	4.55%	90.91%
Architecture, Fine and Visual Arts	0.00%	0.00%	100.00%
History	0.00%	7.14%	92.86%
English, Literature & Languages	4.08%	4.08%	91.84%
Communications & Journalism	0.00%	0.00%	100.00%
Engineering	18.92%	5.41%	75.68%
Philosophy & Religion	0.00%	16.67%	83.33%
Economics/Finance	11.11%	0.00%	88.89%
Business (other than economics/finance)	10.77%	6.15%	83.08%
Performing Arts	0.00%	0.00%	100.00%
Psychology & Social Work	9.68%	0.00%	90.32%
Medicine	10.26%	7.69%	82.05%
Law	16.67%	0.00%	83.33%
Mathematics/Statistics/Computer Science	5.71%	0.00%	94.29%
Hard Sciences (Chemistry, Biology, Physics)	7.77%	4.85%	87.38%
Other	0.00%	25.00%	75.00%

Table 2.12 Have you submitted any of your articles, data, lab logs, code, blog posts, class records or videos, personally developed educational materials or any other content connected to your research for inclusion in an AI model? Broken out by years at present institution

Years at present institution	No Answer	Yes	No
3 years or less	12.50%	0.00%	87.50%
3+ years to 5 years	3.57%	7.14%	89.29%
5+ years to 10 years	5.19%	5.19%	89.61%
10+ years to 20 years	7.46%	3.98%	88.56%
More than 20 years	7.96%	4.78%	87.26%

If you have already contributed content to train an AI application briefly describe what you have contributed and, if possible, to what kind of effort.

Social Sciences (except economics and psychology)

1 If I have never used AI as indicated in the first few questions, there should have been an option to skip to a question # that is actually relevant to me.
2 I have in academic administration, helped to train an AI chatbot for recruitment.
3 No
4 I am hesitant to use AI in ways that I might benefit from because the various services will not say what sources they are using 'to train on.'

Economics/Finance

1 I put almost all my class materials up on my blog. That should be available to AIs for training simply as part of the internet. That is the only thing I have done to "ready and archive" those materials.

Business (other than economics/finance)

1 None
2 I have uploaded papers to ChatGPT to ask it for feedback on structure. I have not knowingly provided content to train anything but I know my students have uploaded my proprietary materials to ChatGPT repeatedly without my permission
3 I have done this, but I use AI Agents (You.com and Co-Pilot) which do not save the content or output to its own archive.
4 No
5 ACADEMIA.edu created AI-generated podcast of my material/papers, outside of their Terms of Service and without permissions to create derivative works. Many of my publications would appear to have been sucked into AI models in their training,
6 No and I'm not sure that I would.
7 Nothing
8 Used a large corpus of text documents that I collected to train a BERT model first, and then a Word2Vec model after the BERT model failed.
9 No
10 I assume all of my public-visible content has been used to train one or more AIs
11 None.
12 An entire book manuscript

Psychology & Social Work

1 I have course material and articles out there, but I don't consent to them being used in AI training.
2 LARGE DATA SETS

Medicine

1 I fed the AI my course objectives and trusted resources on how to communicate and trained the ai to be a simulated client for my students to train their comms skills
2 None
3 May be YouTube videos?
4 A data set about clinical staging of a medical condition
5 No
6 Within CoPilot, I have done some training to create consistent student evaluations, as well as consistent reviews/summaries of applicant materials.
7 How to organize an electronic lab notebook for protocol description.

Mathematics/Statistics/Computer Science

1 My dissertation (20 years ago) on gender equity in Computer Science could be used to train an AI model for more female inclusivity in a male dominated field.
2 Medical outcomes data

Hard Sciences (Chemistry, Biology, Physics)

1 Image analysis. Didn't work well.
2 No
3 AI has already stolen my content from illegal sources.
4 My published articles and chapters appear in ChatGPT search, but I've not personally submitted any work for inclusion in an AI model.
5 None
6 We develop and train our own models using our own data.
7 We use experimental data to train deep reasoning nets that interpret the experiments
8 No
9 Image analysis (of microscopy images)
10 AI is not useful for my research
11 I am sure my work has been stolen by AI companies to train their models. I therefore consider AI companies to be criminal and nature and stay away from such criminal enterprises.
12 Some educational outcome data. Also some literature search data for binding affinities.
13 Movement sounds of pollinators in fruit crops

Education or Library Science

1 Training the AI with more accurate local information for public libraries
2 Creating editing suggestions and feedback on papers.
3 Not applicable
4 No

Allied Health: Nursing, Occupation or Physical Therapy, Nutrition, etc.

1 I have uploaded transcripts of interviews and previous non-peer-reviewed articles to help with writing a new non-peer-reviewed article. This was using a Yale-specific garden-walled version of ChatGPT.
2 No
3 Isn't everything publicly available used to train AI models?

Earth and Environmental Sciences

1 I have not used AI. I have been author or co-author on 86 published papers. According to Academia, as of 6/24/25, my name has been mentioned in 26,286 papers.
2 I am the chairs and lead conveners of a few meeting organizations and sessions on the topic of AI. I have published many papers that uses AI to solve science questions in (space) physics. I have trained US work force on AI (undergrad, grad, postdocs, research scientists).

History

1 I have an appendix to my book (for which I have recently signed a contract with MIT Press) where I discussed machine learning and how it is different from what I did in the book manuscript. However, I do not consider myself to be an expert in machine learning. I have a little knowledge about it.

English, Literature & Languages

1 In my personal life, I created a custom GPT for a local animal rescue to utilize when writing bios for newly arrived animals. I trained the system using hand selected human-composed bios. I have not trained any AI academically.
2 See previous cranky post. To my knowledge, little of my work has been stolen to train AI, thankfully. The word "content" is itself a hateful construct. That's my thinking, my world, my brain. It is not lorem ipsum.
3 An article at publisher's request--no idea the effect.
4 According to a database published in _The Atlantic_ earlier this year Meta has already used my first book (published by the University of Michigan Press) without my or my publisher's permission. So I have contributed content, but without my consent (and over my strenuous objections.)

5 I involuntarily "contributed" all my published writing to this effort via Meta's bot (as did my dad and grandfather, whose scholarly work is also included in their data dump)
6 articles
7 I have generated prompts and evaluated answers.

Communications & Journalism

1 Not yet - my dissertation will involve AI models that can develop customized learning plans that will assist with vocational training certification exam passing.
2 I'm sure my content has been used against my will, as my content is routinely stolen by students and uploaded to Chegg, Course Hero, etc. I have found my stolen content re-posted and used in other schools too.

Engineering

1 Automated vision recognition systems
2 No
3 I only shared my experienced with my peers and within dedicated workshop activities that has blog style communication.

Philosophy & Religion

1 I gave a publisher permission to do with work I had published with them.
2 I am opposed to contributing content to train AI.

Table 3 To which of the following have you ever contributed your content outlined in the past two questions for the development of an AI model by any of the parties below? (Check all that apply)

Table 3.1.1 To which of the following have you ever contributed your content outlined in the past two questions for the development of an AI model by any of the parties below? Private Tech Company Developing an AI Model.

	No Answer	Yes	No
Entire sample	0.00%	1.97%	98.03%

Table 3.1.2 To which of the following have you ever contributed your content outlined in the past two questions for the development of an AI model by any of the parties below? Private Tech Company Developing an AI Model. Broken out by enrollment

Enrollment	Yes	No
Less than 5,000	0.60%	99.40%
5,000+ - 17,000	2.40%	97.60%
17,000+ - 32,000	2.38%	97.62%
More than 32,000	2.55%	97.45%

Table 3.1.3 To which of the following have you ever contributed your content outlined in the past two questions for the development of an AI model by any of the parties below? Private Tech Company Developing an AI Model. Broken out by type of college or Carnegie Class

Type of college or Carnegie Class	Yes	No
Community College	0.00%	100.00%
BA-Granting Institution	0.00%	100.00%
Specialized College (i.e. performing arts, seminary, etc.)	5.56%	94.44%
MA-PHD Granting College	2.26%	97.74%
Level 1 or Level 2 Research University	2.16%	97.84%

Table 3.1.4 To which of the following have you ever contributed your content outlined in the past two questions for the development of an AI model by any of the parties below? Private Tech Company Developing an AI Model. Broken out by public or private college

Public or private college	Yes	No
Public	2.42%	97.58%
Private	0.97%	99.03%

Table 3.1.5 To which of the following have you ever contributed your content outlined in the past two questions for the development of an AI model by any of the parties below? Private Tech Company Developing an AI Model. Broken out by age of respondent

age of respondent	Yes	No
30 or under	11.11%	88.89%
31-39	2.50%	97.50%
40-49	1.94%	98.06%
50-59	2.58%	97.42%
60 or over	1.44%	98.56%
Unknown	0.00%	100.00%

Table 3.1.6 To which of the following have you ever contributed your content outlined in the past two questions for the development of an AI model by any of the parties below? Private Tech Company Developing an AI Model. Broken out by political views

Political views	Yes	No
Left of Liberal	2.97%	97.03%
Liberal	1.18%	98.82%
Middle of the Road	2.07%	97.93%
Conservative	2.00%	98.00%
Unknown	9.09%	90.91%

Table 3.1.7 To which of the following have you ever contributed your content outlined in the past two questions for the development of an AI model by any of the parties below? Private Tech Company Developing an AI Model. Broken out by annual personal income

annual personal income	Yes	No
Less than $50,000	0.00%	100.00%
$50,000 to $75,000	3.08%	96.92%
$75,000+ to $100,000	2.59%	97.41%
$100,000 to $150,000	0.00%	100.00%
More than $150,000	3.24%	96.76%
Unknown	0.00%	100.00%

Table 3.1.8 To which of the following have you ever contributed your content outlined in the past two questions for the development of an AI model by any of the parties below? Private Tech Company Developing an AI Model. Broken out by tenure track

Tenure track	Yes	No
Tenured	2.06%	97.94%
Not tenured but on a tenure track	2.70%	97.30%
Not tenured and not on a tenure track	1.73%	98.27%
Unknown	0.00%	100.00%

Table 3.1.9 To which of the following have you ever contributed your content outlined in the past two questions for the development of an AI model by any of the parties below? Private Tech Company Developing an AI Model. Broken out by gender of respondent

Gender of respondent	Yes	No
Male	2.21%	97.79%
Female	1.86%	98.14%
Transgender	0.00%	100.00%
Unknown	0.00%	100.00%

Table 3.1.10 To which of the following have you ever contributed your content outlined in the past two questions for the development of an AI model by any of the parties below? Private Tech Company Developing an AI Model. Broken out by race or ethnicity

Race or ethnicity	Yes	No
White or Caucasian	1.49%	98.51%
Black or African American	11.54%	88.46%
Hispanic or Latino	0.00%	100.00%
Asian or Asian American	2.63%	97.37%
Mixed Race	5.56%	94.44%
Unknown	0.00%	100.00%

Table 3.1.11 To which of the following have you ever contributed your content outlined in the past two questions for the development of an AI model by any of the parties below? Private Tech Company Developing an AI Model. Broken out by academic field

Academic field	Yes	No
Social Sciences (except economics and psychology)	0.00%	100.00%
Education or Library Science	0.00%	100.00%
Allied Health: Nursing, Occupation or Physical Therapy, Nutrition, etc.	0.00%	100.00%
Earth and Environmental Sciences	0.00%	100.00%
Architecture, Fine and Visual Arts	0.00%	100.00%
History	0.00%	100.00%
English, Literature & Languages	6.12%	93.88%
Communications & Journalism	0.00%	100.00%
Engineering	2.70%	97.30%
Philosophy & Religion	0.00%	100.00%
Economics/Finance	0.00%	100.00%
Business (other than economics/finance)	6.15%	93.85%
Performing Arts	0.00%	100.00%
Psychology & Social Work	3.23%	96.77%
Medicine	3.85%	96.15%
Law	0.00%	100.00%
Mathematics/Statistics/Computer Science	0.00%	100.00%
Hard Sciences (Chemistry, Biology, Physics)	0.00%	100.00%
Other	25.00%	75.00%

Table 3.1.12 To which of the following have you ever contributed your content outlined in the past two questions for the development of an AI model by any of the parties below? Private Tech Company Developing an AI Model. Broken out by years at present institution

Years at present institution	Yes	No
3 years or less	7.50%	92.50%
3+ years to 5 years	7.14%	92.86%
5+ years to 10 years	1.30%	98.70%
10+ years to 20 years	1.00%	99.00%
More than 20 years	1.59%	98.41%

Table 3.2.1 To which of the following have you ever contributed your content outlined in the past two questions for the development of an AI model by any of the parties below? Journal, Book or Other Content Publisher.

	No Answer	Yes	No
Entire sample	0.00%	4.39%	95.61%

Table 3.2.2 To which of the following have you ever contributed your content outlined in the past two questions for the development of an AI model by any of the parties below? Journal, Book or Other Content Publisher. Broken out by enrollment

Enrollment	Yes	No
Less than 5,000	4.76%	95.24%
5,000+ - 17,000	5.39%	94.61%
17,000+ - 32,000	2.98%	97.02%
More than 32,000	4.46%	95.54%

Table 3.2.3 To which of the following have you ever contributed your content outlined in the past two questions for the development of an AI model by any of the parties below? Journal, Book or Other Content Publisher. Broken out by type of college or Carnegie Class

Type of college or Carnegie Class	Yes	No
Community College	0.00%	100.00%
BA-Granting Institution	1.82%	98.18%
Specialized College (i.e. performing arts, seminary, etc.)	0.00%	100.00%
MA-PHD Granting College	3.01%	96.99%
Level 1 or Level 2 Research University	5.76%	94.24%

Table 3.2.4 To which of the following have you ever contributed your content outlined in the past two questions for the development of an AI model by any of the parties below? Journal, Book or Other Content Publisher. Broken out by public or private college

Public or private college	Yes	No
Public	4.63%	95.37%
Private	3.88%	96.12%

Table 3.2.5 To which of the following have you ever contributed your content outlined in the past two questions for the development of an AI model by any of the parties below? Journal, Book or Other Content Publisher. Broken out by age of respondent

Age of respondent	Yes	No
30 or under	0.00%	100.00%
31-39	0.00%	100.00%
40-49	6.80%	93.20%
50-59	1.94%	98.06%
60 or over	5.48%	94.52%
Unknown	0.00%	100.00%

Table 3.2.6 To which of the following have you ever contributed your content outlined in the past two questions for the development of an AI model by any of the parties below? Journal, Book or Other Content Publisher. Broken out by political views

Political views	Yes	No
Left of Liberal	0.99%	99.01%
Liberal	3.92%	96.08%
Middle of the Road	7.25%	92.75%
Conservative	3.00%	97.00%
Unknown	9.09%	90.91%

Table 3.2.7 To which of the following have you ever contributed your content outlined in the past two questions for the development of an AI model by any of the parties below? Journal, Book or Other Content Publisher. Broken out by annual personal income

Annual personal income	Yes	No
Less than $50,000	0.00%	100.00%
$50,000 to $75,000	3.08%	96.92%
$75,000+ to $100,000	2.59%	97.41%
$100,000 to $150,000	2.49%	97.51%
More than $150,000	7.69%	92.31%
Unknown	0.00%	100.00%

Table 3.2.8 To which of the following have you ever contributed your content outlined in the past two questions for the development of an AI model by any of the parties below? Journal, Book or Other Content Publisher. Broken out by tenure track

Tenure track	Yes	No
Tenured	4.64%	95.36%
Not tenured but on a tenure track	5.41%	94.59%
Not tenured and not on a tenure track	3.90%	96.10%
Unknown	0.00%	100.00%

Table 3.2.9 To which of the following have you ever contributed your content outlined in the past two questions for the development of an AI model by any of the parties below? Journal, Book or Other Content Publisher. Broken out by gender of respondent

Gender of respondent	Yes	No
Male	3.31%	96.69%
Female	5.32%	94.68%
Transgender	0.00%	100.00%
Unknown	0.00%	100.00%

Table 3.2.10 To which of the following have you ever contributed your content outlined in the past two questions for the development of an AI model by any of the parties below? Journal, Book or Other Content Publisher. Broken out by race or ethnicity

Race or ethnicity	Yes	No
White or Caucasian	4.85%	95.15%
Black or African American	0.00%	100.00%
Hispanic or Latino	0.00%	100.00%
Asian or Asian American	7.89%	92.11%
Mixed Race	0.00%	100.00%
Unknown	0.00%	100.00%

Table 3.2.11 To which of the following have you ever contributed your content outlined in the past two questions for the development of an AI model by any of the parties below? Journal, Book or Other Content Publisher. Broken out by academic field

Academic field	Yes	No
Social Sciences (except economics and psychology)	0.00%	100.00%
Education or Library Science	4.76%	95.24%
Allied Health: Nursing, Occupation or Physical Therapy, Nutrition, etc.	0.00%	100.00%
Earth and Environmental Sciences	9.09%	90.91%
Architecture, Fine and Visual Arts	0.00%	100.00%
History	0.00%	100.00%
English, Literature & Languages	6.12%	93.88%
Communications & Journalism	0.00%	100.00%
Engineering	8.11%	91.89%
Philosophy & Religion	16.67%	83.33%
Economics/Finance	0.00%	100.00%
Business (other than economics/finance)	4.62%	95.38%
Performing Arts	5.00%	95.00%
Psychology & Social Work	0.00%	100.00%
Medicine	6.41%	93.59%
Law	16.67%	83.33%
Mathematics/Statistics/Computer Science	2.86%	97.14%
Hard Sciences (Chemistry, Biology, Physics)	5.83%	94.17%
Other	25.00%	75.00%

Table 3.2.12 To which of the following have you ever contributed your content outlined in the past two questions for the development of an AI model by any of the parties below? Journal, Book or Other Content Publisher. Broken out by years at present institution

Years at present institution	Yes	No
3 years or less	0.00%	100.00%
3+ years to 5 years	3.57%	96.43%
5+ years to 10 years	9.09%	90.91%
10+ years to 20 years	2.99%	97.01%
More than 20 years	4.78%	95.22%

Table 3.3.1 To which of the following have you ever contributed your content outlined in the past two questions for the development of an AI model by any of the parties below? Peers in my or Related Field.

	No Answer	Yes	No
Entire sample	0.00%	4.85%	95.15%

Table 3.3.2 To which of the following have you ever contributed your content outlined in the past two questions for the development of an AI model by any of the parties below? Peers in my or Related Field. Broken out by enrollment

Enrollment	Yes	No
Less than 5,000	4.17%	95.83%
5,000+ - 17,000	4.19%	95.81%
17,000+ - 32,000	7.74%	92.26%
More than 32,000	3.18%	96.82%

Table 3.3.3 To which of the following have you ever contributed your content outlined in the past two questions for the development of an AI model by any of the parties below? Peers in my or Related Field. Broken out by type of college or Carnegie Class

Type of college or Carnegie Class	Yes	No
Community College	2.70%	97.30%
BA-Granting Institution	1.82%	98.18%
Specialized College (i.e. performing arts, seminary, etc.)	5.56%	94.44%
MA-PHD Granting College	6.77%	93.23%
Level 1 or Level 2 Research University	4.80%	95.20%

Table 3.3.4 To which of the following have you ever contributed your content outlined in the past two questions for the development of an AI model by any of the parties below? Peers in my or Related Field. Broken out by public or private college

Public or private college	Yes	No
Public	5.07%	94.93%
Private	4.37%	95.63%

Table 3.3.5 To which of the following have you ever contributed your content outlined in the past two questions for the development of an AI model by any of the parties below? Peers in my or Related Field. Broken out by age of respondent

Age of respondent	Yes	No
30 or under	0.00%	100.00%
31-39	2.50%	97.50%
40-49	3.88%	96.12%
50-59	5.16%	94.84%
60 or over	5.19%	94.81%
Unknown	16.67%	83.33%

Table 3.3.6 To which of the following have you ever contributed your content outlined in the past two questions for the development of an AI model by any of the parties below? Peers in my or Related Field. Broken out by political views

Political views	Yes	No
Left of Liberal	1.98%	98.02%
Liberal	4.31%	95.69%
Middle of the Road	4.66%	95.34%
Conservative	9.00%	91.00%
Unknown	9.09%	90.91%

Table 3.3.7 To which of the following have you ever contributed your content outlined in the past two questions for the development of an AI model by any of the parties below? Peers in my or Related Field. Broken out by annual personal income

Annual personal income	Yes	No
Less than $50,000	12.50%	87.50%
$50,000 to $75,000	4.62%	95.38%
$75,000+ to $100,000	2.59%	97.41%
$100,000 to $150,000	5.47%	94.53%
More than $150,000	5.26%	94.74%
Unknown	0.00%	100.00%

Table 3.3.8 To which of the following have you ever contributed your content outlined in the past two questions for the development of an AI model by any of the parties below? Peers in my or Related Field. Broken out by tenure track

Tenure track	Yes	No
Tenured	5.15%	94.85%
Not tenured but on a tenure track	2.70%	97.30%
Not tenured and not on a tenure track	4.76%	95.24%
Unknown	0.00%	100.00%

Table 3.3.9 To which of the following have you ever contributed your content outlined in the past two questions for the development of an AI model by any of the parties below? Peers in my or Related Field. Broken out by gender of respondent

Gender of respondent	Yes	No
Male	5.51%	94.49%
Female	4.52%	95.48%
Transgender	0.00%	100.00%
Unknown	0.00%	100.00%

Table 3.3.10 To which of the following have you ever contributed your content outlined in the past two questions for the development of an AI model by any of the parties below? Peers in my or Related Field. Broken out by race or ethnicity

Race or ethnicity	Yes	No
White or Caucasian	4.29%	95.71%
Black or African American	7.69%	92.31%
Hispanic or Latino	11.11%	88.89%
Asian or Asian American	10.53%	89.47%
Mixed Race	0.00%	100.00%
Unknown	0.00%	100.00%

Table 3.3.11 To which of the following have you ever contributed your content outlined in the past two questions for the development of an AI model by any of the parties below? Peers in my or Related Field. Broken out by academic field

Academic field	Yes	No
Social Sciences (except economics and psychology)	3.08%	96.92%
Education or Library Science	9.52%	90.48%
Allied Health: Nursing, Occupation or Physical Therapy, Nutrition, etc.	2.27%	97.73%
Earth and Environmental Sciences	4.55%	95.45%
Architecture, Fine and Visual Arts	10.00%	90.00%
History	0.00%	100.00%
English, Literature & Languages	4.08%	95.92%
Communications & Journalism	9.09%	90.91%
Engineering	13.51%	86.49%
Philosophy & Religion	0.00%	100.00%
Economics/Finance	0.00%	100.00%
Business (other than economics/finance)	7.69%	92.31%
Performing Arts	0.00%	100.00%
Psychology & Social Work	6.45%	93.55%
Medicine	1.28%	98.72%
Law	0.00%	100.00%
Mathematics/Statistics/Computer Science	8.57%	91.43%
Hard Sciences (Chemistry, Biology, Physics)	2.91%	97.09%
Other	25.00%	75.00%

Table 3.3.12 To which of the following have you ever contributed your content outlined in the past two questions for the development of an AI model by any of the parties below? Peers in my or Related Field. Broken out by years at present institution

Years at present institution	Yes	No
3 years or less	2.50%	97.50%
3+ years to 5 years	3.57%	96.43%
5+ years to 10 years	10.39%	89.61%
10+ years to 20 years	2.49%	97.51%
More than 20 years	5.41%	94.59%

Table 3.4.1 To which of the following have you ever contributed your content outlined in the past two questions for the development of an AI model by any of the parties below? Model from Trade or Professional Association.

	No Answer	Yes	No
Entire sample	0.00%	1.21%	98.79%

Table 3.4.2 To which of the following have you ever contributed your content outlined in the past two questions for the development of an AI model by any of the parties below? Model from Trade or Professional Association. Broken out by enrollment

Enrollment	Yes	No
Less than 5,000	2.98%	97.02%
5,000+ - 17,000	0.60%	99.40%
17,000+ - 32,000	0.60%	99.40%
More than 32,000	0.64%	99.36%

Table 3.4.3 To which of the following have you ever contributed your content outlined in the past two questions for the development of an AI model by any of the parties below? Model from Trade or Professional Association. Broken out by type of college or Carnegie Class

Type of college or Carnegie Class	Yes	No
Community College	2.70%	97.30%
BA-Granting Institution	3.64%	96.36%
Specialized College (i.e. performing arts, seminary, etc.)	5.56%	94.44%
MA-PHD Granting College	0.75%	99.25%
Level 1 or Level 2 Research University	0.72%	99.28%

Table 3.4.4 To which of the following have you ever contributed your content outlined in the past two questions for the development of an AI model by any of the parties below? Model from Trade or Professional Association. Broken out by public or private college

Public or private college	Yes	No
Public	1.10%	98.90%
Private	1.46%	98.54%

Table 3.4.5 To which of the following have you ever contributed your content outlined in the past two questions for the development of an AI model by any of the parties below? Model from Trade or Professional Association. Broken out by age of respondent

Age of respondent	Yes	No
30 or under	0.00%	100.00%
31-39	0.00%	100.00%
40-49	0.00%	100.00%
50-59	2.58%	97.42%
60 or over	1.15%	98.85%
Unknown	0.00%	100.00%

Table 3.4.6 To which of the following have you ever contributed your content outlined in the past two questions for the development of an AI model by any of the parties below? Model from Trade or Professional Association. Broken out by political views

Political views	Yes	No
Left of Liberal	1.98%	98.02%
Liberal	0.39%	99.61%
Middle of the Road	1.04%	98.96%
Conservative	3.00%	97.00%
Unknown	0.00%	100.00%

Table 3.4.7 To which of the following have you ever contributed your content outlined in the past two questions for the development of an AI model by any of the parties below? Model from Trade or Professional Association. Broken out by annual personal income

Annual personal income	Yes	No
Less than $50,000	0.00%	100.00%
$50,000 to $75,000	3.08%	96.92%
$75,000+ to $100,000	1.72%	98.28%
$100,000 to $150,000	1.00%	99.00%
More than $150,000	0.81%	99.19%
Unknown	0.00%	100.00%

Table 3.4.8 To which of the following have you ever contributed your content outlined in the past two questions for the development of an AI model by any of the parties below? Model from Trade or Professional Association. Broken out by tenure track

Tenure track	Yes	No
Tenured	1.03%	98.97%
Not tenured but on a tenure track	0.00%	100.00%
Not tenured and not on a tenure track	1.73%	98.27%
Unknown	0.00%	100.00%

Table 3.4.9 To which of the following have you ever contributed your content outlined in the past two questions for the development of an AI model by any of the parties below? Model from Trade or Professional Association. Broken out by gender of respondent

Gender of respondent	Yes	No
Male	1.47%	98.53%
Female	1.06%	98.94%
Transgender	0.00%	100.00%
Unknown	0.00%	100.00%

Table 3.4.10 To which of the following have you ever contributed your content outlined in the past two questions for the development of an AI model by any of the parties below? Model from Trade or Professional Association. Broken out by race or ethnicity

Race or ethnicity	Yes	No
White or Caucasian	0.93%	99.07%
Black or African American	3.85%	96.15%
Hispanic or Latino	3.70%	96.30%
Asian or Asian American	2.63%	97.37%
Mixed Race	0.00%	100.00%
Unknown	0.00%	100.00%

Table 3.4.11 To which of the following have you ever contributed your content outlined in the past two questions for the development of an AI model by any of the parties below? Model from Trade or Professional Association. Broken out by academic field

Academic field	Yes	No
Social Sciences (except economics and psychology)	1.54%	98.46%
Education or Library Science	2.38%	97.62%
Allied Health: Nursing, Occupation or Physical Therapy, Nutrition, etc.	0.00%	100.00%
Earth and Environmental Sciences	0.00%	100.00%
Architecture, Fine and Visual Arts	0.00%	100.00%
History	0.00%	100.00%
English, Literature & Languages	0.00%	100.00%
Communications & Journalism	0.00%	100.00%
Engineering	0.00%	100.00%
Philosophy & Religion	0.00%	100.00%
Economics/Finance	5.56%	94.44%
Business (other than economics/finance)	3.08%	96.92%
Performing Arts	0.00%	100.00%
Psychology & Social Work	9.68%	90.32%
Medicine	0.00%	100.00%
Law	0.00%	100.00%
Mathematics/Statistics/Computer Science	0.00%	100.00%
Hard Sciences (Chemistry, Biology, Physics)	0.00%	100.00%
Other	0.00%	100.00%

Table 3.4.12 To which of the following have you ever contributed your content outlined in the past two questions for the development of an AI model by any of the parties below? Model from Trade or Professional Association. Broken out by years at present institution

Years at present institution	Yes	No
3 years or less	2.50%	97.50%
3+ years to 5 years	0.00%	100.00%
5+ years to 10 years	1.30%	98.70%
10+ years to 20 years	1.49%	98.51%
More than 20 years	0.96%	99.04%

Table 3.5.1 To which of the following have you ever contributed your content outlined in the past two questions for the development of an AI model by any of the parties below? University Developed Model.

	No Answer	Yes	No
Entire sample	0.00%	3.18%	96.82%

Table 3.5.2 To which of the following have you ever contributed your content outlined in the past two questions for the development of an AI model by any of the parties below? University Developed Model. Broken out by enrollment

Enrollment	Yes	No
Less than 5,000	3.57%	96.43%
5,000+ - 17,000	3.59%	96.41%
17,000+ - 32,000	1.79%	98.21%
More than 32,000	3.82%	96.18%

Table 3.5.3 To which of the following have you ever contributed your content outlined in the past two questions for the development of an AI model by any of the parties below? University Developed Model. Broken out by type of college or Carnegie Class

Type of college or Carnegie Class	Yes	No
Community College	0.00%	100.00%
BA-Granting Institution	3.64%	96.36%
Specialized College (i.e. performing arts, seminary, etc.)	5.56%	94.44%
MA-PHD Granting College	3.76%	96.24%
Level 1 or Level 2 Research University	3.12%	96.88%

Table 3.5.4 To which of the following have you ever contributed your content outlined in the past two questions for the development of an AI model by any of the parties below? University Developed Model. Broken out by public or private college

Public or private college	Yes	No
Public	2.86%	97.14%
Private	3.88%	96.12%

Table 3.5.5 To which of the following have you ever contributed your content outlined in the past two questions for the development of an AI model by any of the parties below? University Developed Model. Broken out by age of respondent

Age of respondent	Yes	No
30 or under	0.00%	100.00%
31-39	0.00%	100.00%
40-49	3.88%	96.12%
50-59	1.94%	98.06%
60 or over	3.75%	96.25%
Unknown	16.67%	83.33%

Table 3.5.6 To which of the following have you ever contributed your content outlined in the past two questions for the development of an AI model by any of the parties below? University Developed Model. Broken out by political views

Political views	Yes	No
Left of Liberal	0.00%	100.00%
Liberal	3.53%	96.47%
Middle of the Road	2.07%	97.93%
Conservative	8.00%	92.00%
Unknown	0.00%	100.00%

Table 3.5.7 To which of the following have you ever contributed your content outlined in the past two questions for the development of an AI model by any of the parties below? University Developed Model. Broken out by annual personal income

Annual personal income	Yes	No
Less than $50,000	0.00%	100.00%
$50,000 to $75,000	4.62%	95.38%
$75,000+ to $100,000	1.72%	98.28%
$100,000 to $150,000	4.98%	95.02%
More than $150,000	2.43%	97.57%
Unknown	0.00%	100.00%

Table 3.5.8 To which of the following have you ever contributed your content outlined in the past two questions for the development of an AI model by any of the parties below? University Developed Model. Broken out by tenure track

Tenure track	Yes	No
Tenured	1.55%	98.45%
Not tenured but on a tenure track	5.41%	94.59%
Not tenured and not on a tenure track	5.63%	94.37%
Unknown	0.00%	100.00%

Table 3.5.9 To which of the following have you ever contributed your content outlined in the past two questions for the development of an AI model by any of the parties below? University Developed Model. Broken out by gender of respondent

Gender of respondent	Yes	No
Male	3.31%	96.69%
Female	3.19%	96.81%
Transgender	0.00%	100.00%
Unknown	0.00%	100.00%

Table 3.5.10 To which of the following have you ever contributed your content outlined in the past two questions for the development of an AI model by any of the parties below? University Developed Model. Broken out by race or ethnicity

Race or ethnicity	Yes	No
White or Caucasian	2.99%	97.01%
Black or African American	3.85%	96.15%
Hispanic or Latino	3.70%	96.30%
Asian or Asian American	0.00%	100.00%
Mixed Race	16.67%	83.33%
Unknown	0.00%	100.00%

Table 3.5.11 To which of the following have you ever contributed your content outlined in the past two questions for the development of an AI model by any of the parties below? University Developed Model. Broken out by academic field

Academic field	Yes	No
Social Sciences (except economics and psychology)	1.54%	98.46%
Education or Library Science	2.38%	97.62%
Allied Health: Nursing, Occupation or Physical Therapy, Nutrition, etc.	6.82%	93.18%
Earth and Environmental Sciences	0.00%	100.00%
Architecture, Fine and Visual Arts	0.00%	100.00%
History	0.00%	100.00%
English, Literature & Languages	2.04%	97.96%
Communications & Journalism	0.00%	100.00%
Engineering	8.11%	91.89%
Philosophy & Religion	0.00%	100.00%
Economics/Finance	5.56%	94.44%
Business (other than economics/finance)	7.69%	92.31%
Performing Arts	0.00%	100.00%
Psychology & Social Work	0.00%	100.00%
Medicine	3.85%	96.15%
Law	0.00%	100.00%
Mathematics/Statistics/Computer Science	2.86%	97.14%
Hard Sciences (Chemistry, Biology, Physics)	1.94%	98.06%
Other	0.00%	100.00%

Table 3.5.12 To which of the following have you ever contributed your content outlined in the past two questions for the development of an AI model by any of the parties below? University Developed Model. Broken out by years at present institution

Years at present institution	Yes	No
3 years or less	2.50%	97.50%
3+ years to 5 years	3.57%	96.43%
5+ years to 10 years	1.30%	98.70%
10+ years to 20 years	2.99%	97.01%
More than 20 years	3.82%	96.18%

Table 4.1 To the best of your knowledge has content developed by you been used by an AI model without your permission?

	No Answer	Yes	No	Maybe but I am not sure
Entire sample	9.85%	9.24%	40.61%	40.30%

Table 4.2 To the best of your knowledge has content developed by you been used by an AI model without your permission? Broken out by enrollment

Enrollment	No Answer	Yes	No	Maybe but I am not sure
Less than 5,000	9.52%	7.74%	42.26%	40.48%
5,000+ - 17,000	7.78%	8.38%	38.92%	44.91%
17,000+ - 32,000	10.71%	10.71%	41.67%	36.90%
More than 32,000	11.46%	10.19%	39.49%	38.85%

Table 4.3 To the best of your knowledge has content developed by you been used by an AI model without your permission? Broken out by type of college or Carnegie Class

Type of college or Carnegie Class	No Answer	Yes	No	Maybe but I am not sure
Community College	2.70%	2.70%	51.35%	43.24%
BA-Granting Institution	3.64%	9.09%	43.64%	43.64%
Specialized College (i.e. performing arts, seminary, etc.)	16.67%	5.56%	33.33%	44.44%
MA-PHD Granting College	11.28%	7.52%	44.36%	36.84%
Level 1 or Level 2 Research University	10.55%	10.55%	38.37%	40.53%

Table 4.4 To the best of your knowledge has content developed by you been used by an AI model without your permission? Broken out by public or private college

Public or private college	No Answer	Yes	No	Maybe but I am not sure
Public	10.79%	10.13%	40.75%	38.33%
Private	7.77%	7.28%	40.29%	44.66%

Table 4.5 To the best of your knowledge has content developed by you been used by an AI model without your permission? Broken out by age of respondent

Age of respondent	No Answer	Yes	No	Maybe but I am not sure
30 or under	11.11%	0.00%	44.44%	44.44%
31-39	10.00%	10.00%	25.00%	55.00%
40-49	8.74%	10.68%	44.66%	35.92%
50-59	9.68%	9.03%	38.71%	42.58%
60 or over	10.09%	9.22%	42.36%	38.33%
Unknown	16.67%	0.00%	16.67%	66.67%

Table 4.6 To the best of your knowledge has content developed by you been used by an AI model without your permission? Broken out by political views

Political views	No Answer	Yes	No	Maybe but I am not sure
Left of Liberal	4.95%	12.87%	37.62%	44.55%
Liberal	10.20%	8.63%	41.96%	39.22%
Middle of the Road	11.40%	9.33%	38.86%	40.41%
Conservative	12.00%	6.00%	43.00%	39.00%
Unknown	0.00%	18.18%	45.45%	36.36%

Table 4.7 To the best of your knowledge has content developed by you been used by an AI model without your permission? Broken out by annual personal income

Annual personal income	No Answer	Yes	No	Maybe but I am not sure
Less than $50,000	18.75%	0.00%	50.00%	31.25%
$50,000 to $75,000	4.62%	7.69%	50.77%	36.92%
$75,000+ to $100,000	8.62%	11.21%	38.79%	41.38%
$100,000 to $150,000	8.96%	6.97%	43.78%	40.30%
More than $150,000	11.34%	10.93%	38.06%	39.68%
Unknown	20.00%	13.33%	0.00%	66.67%

Table 4.8 To the best of your knowledge has content developed by you been used by an AI model without your permission? Broken out by tenure track

Tenure track	No Answer	Yes	No	Maybe but I am not sure
Tenured	9.02%	11.60%	35.57%	43.81%
Not tenured but on a tenure track	5.41%	8.11%	32.43%	54.05%
Not tenured and not on a tenure track	11.26%	5.63%	50.65%	32.47%
Unknown	50.00%	0.00%	25.00%	25.00%

Table 4.9 To the best of your knowledge has content developed by you been used by an AI model without your permission? Broken out by gender of respondent

Gender of respondent	No Answer	Yes	No	Maybe but I am not sure
Male	10.29%	7.35%	43.38%	38.97%
Female	9.57%	10.64%	38.30%	41.49%
Transgender	0.00%	20.00%	40.00%	40.00%
Unknown	14.29%	0.00%	57.14%	28.57%

Table 4.10 To the best of your knowledge has content developed by you been used by an AI model without your permission? Broken out by race or ethnicity

Race or ethnicity	No Answer	Yes	No	Maybe but I am not sure
White or Caucasian	9.14%	11.01%	40.30%	39.55%
Black or African American	11.54%	0.00%	46.15%	42.31%
Hispanic or Latino	11.11%	0.00%	40.74%	48.15%
Asian or Asian American	21.05%	0.00%	52.63%	26.32%
Mixed Race	0.00%	11.11%	22.22%	66.67%
Unknown	13.33%	0.00%	33.33%	53.33%

Table 4.11 To the best of your knowledge has content developed by you been used by an AI model without your permission? Broken out by academic field

Academic field	No Answer	Yes	No	Maybe but I am not sure
Social Sciences (except economics and psychology)	6.15%	15.38%	33.85%	44.62%
Education or Library Science	11.90%	7.14%	54.76%	26.19%
Allied Health: Nursing, Occupation or Physical Therapy, Nutrition, etc.	11.36%	4.55%	50.00%	34.09%
Earth and Environmental Sciences	9.09%	9.09%	36.36%	45.45%
Architecture, Fine and Visual Arts	0.00%	0.00%	50.00%	50.00%
History	0.00%	7.14%	42.86%	50.00%
English, Literature & Languages	2.04%	20.41%	32.65%	44.90%
Communications & Journalism	0.00%	27.27%	36.36%	36.36%
Engineering	18.92%	10.81%	29.73%	40.54%
Philosophy & Religion	0.00%	16.67%	33.33%	50.00%
Economics/Finance	16.67%	16.67%	27.78%	38.89%
Business (other than economics/finance)	13.85%	1.54%	50.77%	33.85%
Performing Arts	20.00%	5.00%	45.00%	30.00%
Psychology & Social Work	9.68%	9.68%	38.71%	41.94%
Medicine	10.26%	2.56%	43.59%	43.59%
Law	16.67%	0.00%	16.67%	66.67%
Mathematics/Statistics/Computer Science	8.57%	17.14%	42.86%	31.43%
Hard Sciences (Chemistry, Biology, Physics)	9.71%	8.74%	37.86%	43.69%
Other	0.00%	0.00%	25.00%	75.00%

Table 4.12 To the best of your knowledge has content developed by you been used by an AI model without your permission? Broken out by years at present institution

Years at present institution	No Answer	Yes	No	Maybe but I am not sure
3 years or less	15.00%	15.00%	35.00%	35.00%
3+ years to 5 years	7.14%	3.57%	50.00%	39.29%
5+ years to 10 years	7.79%	9.09%	41.56%	41.56%
10+ years to 20 years	7.96%	8.46%	39.80%	43.78%
More than 20 years	11.15%	9.55%	40.76%	38.54%

Table 5.1 Do you know anyone who feels that their content may have been used for an AI model without their permission?

	No Answer	Yes	No
Entire sample	10.45%	25.61%	63.94%

Table 5.2 Do you know anyone who feels that their content may have been used for an AI model without their permission? Broken out by enrollment

Enrollment	No Answer	Yes	No
Less than 5,000	8.93%	23.81%	67.26%
5,000+ - 17,000	9.58%	26.35%	64.07%
17,000+ - 32,000	11.31%	27.98%	60.71%
More than 32,000	12.10%	24.20%	63.69%

Table 5.3 Do you know anyone who feels that their content may have been used for an AI model without their permission? Broken out by type of college or Carnegie Class

Type of college or Carnegie Class	No Answer	Yes	No
Community College	5.41%	24.32%	70.27%
BA-Granting Institution	7.27%	34.55%	58.18%
Specialized College (i.e. performing arts, seminary, etc.)	16.67%	22.22%	61.11%
MA-PHD Granting College	9.02%	27.07%	63.91%
Level 1 or Level 2 Research University	11.51%	24.22%	64.27%

Table 5.4 Do you know anyone who feels that their content may have been used for an AI model without their permission? Broken out by public or private college

Public or private college	No Answer	Yes	No
Public	11.45%	26.43%	62.11%
Private	8.25%	23.79%	67.96%

Table 5.5 Do you know anyone who feels that their content may have been used for an AI model without their permission? Broken out by age of respondent

Age of respondent	No Answer	Yes	No
30 or under	22.22%	11.11%	66.67%
31-39	10.00%	35.00%	55.00%
40-49	12.62%	22.33%	65.05%
50-59	12.26%	30.32%	57.42%
60 or over	8.65%	23.63%	67.72%
Unknown	16.67%	33.33%	50.00%

Table 5.6 Do you know anyone who feels that their content may have been used for an AI model without their permission? Broken out by political views

Political views	No Answer	Yes	No
Left of Liberal	6.93%	33.66%	59.41%
Liberal	10.59%	26.67%	62.75%
Middle of the Road	11.40%	22.28%	66.32%
Conservative	12.00%	23.00%	65.00%
Unknown	9.09%	9.09%	81.82%

Table 5.7 Do you know anyone who feels that their content may have been used for an AI model without their permission? Broken out by annual personal income

Annual personal income	No Answer	Yes	No
Less than $50,000	6.25%	25.00%	68.75%
$50,000 to $75,000	6.15%	29.23%	64.62%
$75,000+ to $100,000	7.76%	28.45%	63.79%
$100,000 to $150,000	10.45%	23.38%	66.17%
More than $150,000	12.15%	25.10%	62.75%
Unknown	26.67%	26.67%	46.67%

Table 5.8 Do you know anyone who feels that their content may have been used for an AI model without their permission? Broken out by tenure track

Tenure track	No Answer	Yes	No
Tenured	10.05%	29.38%	60.57%
Not tenured but on a tenure track	8.11%	27.03%	64.86%
Not tenured and not on a tenure track	10.82%	19.48%	69.70%
Unknown	50.00%	0.00%	50.00%

Table 5.9 Do you know anyone who feels that their content may have been used for an AI model without their permission? Broken out by gender of respondent

Gender of respondent	No Answer	Yes	No
Male	11.40%	24.63%	63.97%
Female	9.57%	26.86%	63.56%
Transgender	20.00%	20.00%	60.00%
Unknown	14.29%	0.00%	85.71%

Table 5.10 Do you know anyone who feels that their content may have been used for an AI model without their permission? Broken out by race or ethnicity

Race or ethnicity	No Answer	Yes	No
White or Caucasian	9.51%	26.87%	63.62%
Black or African American	11.54%	7.69%	80.77%
Hispanic or Latino	11.11%	33.33%	55.56%
Asian or Asian American	23.68%	5.26%	71.05%
Mixed Race	0.00%	50.00%	50.00%
Unknown	20.00%	20.00%	60.00%

Table 5.11 Do you know anyone who feels that their content may have been used for an AI model without their permission? Broken out by academic field

Academic field	No Answer	Yes	No
Social Sciences (except economics and psychology)	6.15%	40.00%	53.85%
Education or Library Science	7.14%	19.05%	73.81%
Allied Health: Nursing, Occupation or Physical Therapy, Nutrition, etc.	11.36%	13.64%	75.00%
Earth and Environmental Sciences	9.09%	22.73%	68.18%
Architecture, Fine and Visual Arts	10.00%	10.00%	80.00%
History	0.00%	14.29%	85.71%
English, Literature & Languages	2.04%	55.10%	42.86%
Communications & Journalism	0.00%	54.55%	45.45%
Engineering	16.22%	18.92%	64.86%
Philosophy & Religion	16.67%	33.33%	50.00%
Economics/Finance	16.67%	27.78%	55.56%
Business (other than economics/finance)	15.38%	18.46%	66.15%
Performing Arts	15.00%	30.00%	55.00%
Psychology & Social Work	16.13%	25.81%	58.06%
Medicine	11.54%	11.54%	76.92%
Law	33.33%	33.33%	33.33%
Mathematics/Statistics/Computer Science	8.57%	31.43%	60.00%
Hard Sciences (Chemistry, Biology, Physics)	9.71%	24.27%	66.02%
Other	25.00%	25.00%	50.00%

Table 5.12 Do you know anyone who feels that their content may have been used for an AI model without their permission? Broken out by years at present institution

Years at present institution	No Answer	Yes	No
3 years or less	17.50%	30.00%	52.50%
3+ years to 5 years	7.14%	25.00%	67.86%
5+ years to 10 years	10.39%	22.08%	67.53%
10+ years to 20 years	9.45%	25.87%	64.68%
More than 20 years	10.51%	25.80%	63.69%

Table 6 Have you personally ever taken part in an effort to develop any of the following?

Table 6.1.1 Have you personally ever taken part in an effort to develop any of the following? Highly detailed prompts for ChatGPT designed as a template to be used by others.

	No Answer	Yes	No
Entire sample	0.00%	6.67%	93.33%

Table 6.1.2 Have you personally ever taken part in an effort to develop any of the following? Highly detailed prompts for ChatGPT designed as a template to be used by others. Broken out by enrollment

Enrollment	Yes	No
Less than 5,000	5.36%	94.64%
5,000+ - 17,000	8.38%	91.62%
17,000+ - 32,000	6.55%	93.45%
More than 32,000	6.37%	93.63%

Table 6.1.3 Have you personally ever taken part in an effort to develop any of the following? Highly detailed prompts for ChatGPT designed as a template to be used by others. Broken out by type of college or Carnegie Class

Type of college or Carnegie Class	Yes	No
Community College	0.00%	100.00%
BA-Granting Institution	7.27%	92.73%
Specialized College (i.e. performing arts, seminary, etc.)	11.11%	88.89%
MA-PHD Granting College	8.27%	91.73%
Level 1 or Level 2 Research University	6.47%	93.53%

Table 6.1.4 Have you personally ever taken part in an effort to develop any of the following? Highly detailed prompts for ChatGPT designed as a template to be used by others. Broken out by public or private college

Public or private college	Yes	No
Public	6.83%	93.17%
Private	6.31%	93.69%

Table 6.1.5 Have you personally ever taken part in an effort to develop any of the following? Highly detailed prompts for ChatGPT designed as a template to be used by others. Broken out by age of respondent

Age of respondent	Yes	No
30 or under	11.11%	88.89%
31-39	10.00%	90.00%
40-49	11.65%	88.35%
50-59	10.32%	89.68%
60 or over	2.88%	97.12%
Unknown	16.67%	83.33%

Table 6.1.6 Have you personally ever taken part in an effort to develop any of the following? Highly detailed prompts for ChatGPT designed as a template to be used by others. Broken out by political views

Political views	Yes	No
Left of Liberal	6.93%	93.07%
Liberal	5.88%	94.12%
Middle of the Road	8.29%	91.71%
Conservative	6.00%	94.00%
Unknown	0.00%	100.00%

Table 6.1.7 Have you personally ever taken part in an effort to develop any of the following? Highly detailed prompts for ChatGPT designed as a template to be used by others. Broken out by annual personal income

Annual personal income	Yes	No
Less than $50,000	6.25%	93.75%
$50,000 to $75,000	10.77%	89.23%
$75,000+ to $100,000	10.34%	89.66%
$100,000 to $150,000	7.46%	92.54%
More than $150,000	3.64%	96.36%
Unknown	0.00%	100.00%

Table 6.1.8 Have you personally ever taken part in an effort to develop any of the following? Highly detailed prompts for ChatGPT designed as a template to be used by others. Broken out by tenure track

Tenure track	Yes	No
Tenured	4.90%	95.10%
Not tenured but on a tenure track	8.11%	91.89%
Not tenured and not on a tenure track	9.52%	90.48%
Unknown	0.00%	100.00%

Table 6.1.9 Have you personally ever taken part in an effort to develop any of the following? Highly detailed prompts for ChatGPT designed as a template to be used by others. Broken out by gender of respondent

Gender of respondent	Yes	No
Male	8.46%	91.54%
Female	5.59%	94.41%
Transgender	0.00%	100.00%
Unknown	0.00%	100.00%

Table 6.1.10 Have you personally ever taken part in an effort to develop any of the following? Highly detailed prompts for ChatGPT designed as a template to be used by others. Broken out by race or ethnicity

Race or ethnicity	Yes	No
White or Caucasian	6.16%	93.84%
Black or African American	7.69%	92.31%
Hispanic or Latino	11.11%	88.89%
Asian or Asian American	13.16%	86.84%
Mixed Race	5.56%	94.44%
Unknown	0.00%	100.00%

Table 6.1.11 Have you personally ever taken part in an effort to develop any of the following? Highly detailed prompts for ChatGPT designed as a template to be used by others. Broken out by academic field

Academic field	Yes	No
Social Sciences (except economics and psychology)	6.15%	93.85%
Education or Library Science	26.19%	73.81%
Allied Health: Nursing, Occupation or Physical Therapy, Nutrition, etc.	6.82%	93.18%
Earth and Environmental Sciences	4.55%	95.45%
Architecture, Fine and Visual Arts	0.00%	100.00%
History	0.00%	100.00%
English, Literature & Languages	12.24%	87.76%
Communications & Journalism	0.00%	100.00%
Engineering	8.11%	91.89%
Philosophy & Religion	0.00%	100.00%
Economics/Finance	11.11%	88.89%
Business (other than economics/finance)	9.23%	90.77%
Performing Arts	10.00%	90.00%
Psychology & Social Work	3.23%	96.77%
Medicine	2.56%	97.44%
Law	0.00%	100.00%
Mathematics/Statistics/Computer Science	0.00%	100.00%
Hard Sciences (Chemistry, Biology, Physics)	2.91%	97.09%
Other	0.00%	100.00%

Table 6.1.12 Have you personally ever taken part in an effort to develop any of the following? Highly detailed prompts for ChatGPT designed as a template to be used by others. Broken out by years at present institution

Years at present institution	Yes	No
3 years or less	10.00%	90.00%
3+ years to 5 years	21.43%	78.57%
5+ years to 10 years	11.69%	88.31%
10+ years to 20 years	8.96%	91.04%
More than 20 years	2.23%	97.77%

Table 6.2.1 Have you personally ever taken part in an effort to develop any of the following? Highly detailed prompts for Claude designed as a template to be used by others.

	No Answer	Yes	No
Entire sample	0.00%	1.67%	98.33%

Table 6.2.2 Have you personally ever taken part in an effort to develop any of the following? Highly detailed prompts for Claude designed as a template to be used by others. Broken out by enrollment

Enrollment	Yes	No
Less than 5,000	1.19%	98.81%
5,000+ - 17,000	1.80%	98.20%
17,000+ - 32,000	2.38%	97.62%
More than 32,000	1.27%	98.73%

Table 6.2.3 Have you personally ever taken part in an effort to develop any of the following? Highly detailed prompts for Claude designed as a template to be used by others. Broken out by type of college or Carnegie Class

Type of college or Carnegie Class	Yes	No
Community College	0.00%	100.00%
BA-Granting Institution	1.82%	98.18%
Specialized College (i.e. performing arts, seminary, etc.)	5.56%	94.44%
MA-PHD Granting College	2.26%	97.74%
Level 1 or Level 2 Research University	1.44%	98.56%

Table 6.2.4 Have you personally ever taken part in an effort to develop any of the following? Highly detailed prompts for Claude designed as a template to be used by others. Broken out by public or private college

Public or private college	Yes	No
Public	1.76%	98.24%
Private	1.46%	98.54%

Table 6.2.5 Have you personally ever taken part in an effort to develop any of the following? Highly detailed prompts for Claude designed as a template to be used by others. Broken out by age of respondent

Age of respondent	Yes	No
30 or under	0.00%	100.00%
31-39	2.50%	97.50%
40-49	5.83%	94.17%
50-59	1.94%	98.06%
60 or over	0.29%	99.71%
Unknown	0.00%	100.00%

Table 6.2.6 Have you personally ever taken part in an effort to develop any of the following? Highly detailed prompts for Claude designed as a template to be used by others. Broken out by political views

Political views	Yes	No
Left of Liberal	0.00%	100.00%
Liberal	2.35%	97.65%
Middle of the Road	2.07%	97.93%
Conservative	1.00%	99.00%
Unknown	0.00%	100.00%

Table 6.2.7 Have you personally ever taken part in an effort to develop any of the following? Highly detailed prompts for Claude designed as a template to be used by others. Broken out by annual personal income

Annual personal income	Yes	No
Less than $50,000	0.00%	100.00%
$50,000 to $75,000	1.54%	98.46%
$75,000+ to $100,000	3.45%	96.55%
$100,000 to $150,000	1.49%	98.51%
More than $150,000	1.21%	98.79%
Unknown	0.00%	100.00%

Table 6.2.8 Have you personally ever taken part in an effort to develop any of the following? Highly detailed prompts for Claude designed as a template to be used by others. Broken out by tenure track

Tenure track	Yes	No
Tenured	1.55%	98.45%
Not tenured but on a tenure track	2.70%	97.30%
Not tenured and not on a tenure track	1.73%	98.27%
Unknown	0.00%	100.00%

Table 6.2.9 Have you personally ever taken part in an effort to develop any of the following? Highly detailed prompts for Claude designed as a template to be used by others. Broken out by gender of respondent

Gender of respondent	Yes	No
Male	2.21%	97.79%
Female	1.33%	98.67%
Transgender	0.00%	100.00%
Unknown	0.00%	100.00%

Table 6.2.10 Have you personally ever taken part in an effort to develop any of the following? Highly detailed prompts for Claude designed as a template to be used by others. Broken out by race or ethnicity

Race or ethnicity	Yes	No
White or Caucasian	1.68%	98.32%
Black or African American	0.00%	100.00%
Hispanic or Latino	3.70%	96.30%
Asian or Asian American	2.63%	97.37%
Mixed Race	0.00%	100.00%
Unknown	0.00%	100.00%

Table 6.2.11 Have you personally ever taken part in an effort to develop any of the following? Highly detailed prompts for Claude designed as a template to be used by others. Broken out by academic field

Academic field	Yes	No
Social Sciences (except economics and psychology)	3.08%	96.92%
Education or Library Science	7.14%	92.86%
Allied Health: Nursing, Occupation or Physical Therapy, Nutrition, etc.	2.27%	97.73%
Earth and Environmental Sciences	4.55%	95.45%
Architecture, Fine and Visual Arts	0.00%	100.00%
History	0.00%	100.00%
English, Literature & Languages	2.04%	97.96%
Communications & Journalism	0.00%	100.00%
Engineering	2.70%	97.30%
Philosophy & Religion	0.00%	100.00%
Economics/Finance	0.00%	100.00%
Business (other than economics/finance)	1.54%	98.46%
Performing Arts	0.00%	100.00%
Psychology & Social Work	3.23%	96.77%
Medicine	0.00%	100.00%
Law	0.00%	100.00%
Mathematics/Statistics/Computer Science	0.00%	100.00%
Hard Sciences (Chemistry, Biology, Physics)	0.00%	100.00%
Other	0.00%	100.00%

Table 6.2.12 Have you personally ever taken part in an effort to develop any of the following? Highly detailed prompts for Claude designed as a template to be used by others. Broken out by years at present institution

Years at present institution	Yes	No
3 years or less	5.00%	95.00%
3+ years to 5 years	3.57%	96.43%
5+ years to 10 years	3.90%	96.10%
10+ years to 20 years	1.99%	98.01%
More than 20 years	0.32%	99.68%

Table 6.3.1 Have you personally ever taken part in an effort to develop any of the following? Highly detailed prompts for Gemini designed as a template to be used by others.

	No Answer	Yes	No
Entire sample	0.00%	1.06%	98.94%

Table 6.3.2 Have you personally ever taken part in an effort to develop any of the following? Highly detailed prompts for Gemini designed as a template to be used by others. Broken out by enrollment

Enrollment	Yes	No
Less than 5,000	0.60%	99.40%
5,000+ - 17,000	0.60%	99.40%
17,000+ - 32,000	1.79%	98.21%
More than 32,000	1.27%	98.73%

Table 6.3.3 Have you personally ever taken part in an effort to develop any of the following? Highly detailed prompts for Gemini designed as a template to be used by others. Broken out by type of college or Carnegie Class

Type of college or Carnegie Class	Yes	No
Community College	0.00%	100.00%
BA-Granting Institution	0.00%	100.00%
Specialized College (i.e. performing arts, seminary, etc.)	5.56%	94.44%
MA-PHD Granting College	1.50%	98.50%
Level 1 or Level 2 Research University	0.96%	99.04%

Table 6.3.4 Have you personally ever taken part in an effort to develop any of the following? Highly detailed prompts for Gemini designed as a template to be used by others. Broken out by public or private college

Public or private college	Yes	No
Public	0.88%	99.12%
Private	1.46%	98.54%

Table 6.3.5 Have you personally ever taken part in an effort to develop any of the following? Highly detailed prompts for Gemini designed as a template to be used by others. Broken out by age of respondent

Age of respondent	Yes	No
30 or under	0.00%	100.00%
31-39	0.00%	100.00%
40-49	3.88%	96.12%
50-59	1.29%	98.71%
60 or over	0.29%	99.71%
Unknown	0.00%	100.00%

Table 6.3.6 Have you personally ever taken part in an effort to develop any of the following? Highly detailed prompts for Gemini designed as a template to be used by others. Broken out by political views

Political views	Yes	No
Left of Liberal	0.00%	100.00%
Liberal	1.18%	98.82%
Middle of the Road	1.04%	98.96%
Conservative	2.00%	98.00%
Unknown	0.00%	100.00%

Table 6.3.7 Have you personally ever taken part in an effort to develop any of the following? Highly detailed prompts for Gemini designed as a template to be used by others. Broken out by annual personal income

Annual personal income	Yes	No
Less than $50,000	0.00%	100.00%
$50,000 to $75,000	0.00%	100.00%
$75,000+ to $100,000	1.72%	98.28%
$100,000 to $150,000	1.99%	98.01%
More than $150,000	0.40%	99.60%
Unknown	0.00%	100.00%

Table 6.3.8 Have you personally ever taken part in an effort to develop any of the following? Highly detailed prompts for Gemini designed as a template to be used by others. Broken out by tenure track

Tenure track	Yes	No
Tenured	1.03%	98.97%
Not tenured but on a tenure track	2.70%	97.30%
Not tenured and not on a tenure track	0.87%	99.13%
Unknown	0.00%	100.00%

Table 6.3.9 Have you personally ever taken part in an effort to develop any of the following? Highly detailed prompts for Gemini designed as a template to be used by others. Broken out by gender of respondent

Gender of respondent	Yes	No
Male	0.74%	99.26%
Female	1.33%	98.67%
Transgender	0.00%	100.00%
Unknown	0.00%	100.00%

Table 6.3.10 Have you personally ever taken part in an effort to develop any of the following? Highly detailed prompts for Gemini designed as a template to be used by others. Broken out by race or ethnicity

Race or ethnicity	Yes	No
White or Caucasian	0.93%	99.07%
Black or African American	0.00%	100.00%
Hispanic or Latino	0.00%	100.00%
Asian or Asian American	5.26%	94.74%
Mixed Race	0.00%	100.00%
Unknown	0.00%	100.00%

Table 6.3.11 Have you personally ever taken part in an effort to develop any of the following? Highly detailed prompts for Gemini designed as a template to be used by others. Broken out by academic field

Academic field	Yes	No
Social Sciences (except economics and psychology)	1.54%	98.46%
Education or Library Science	2.38%	97.62%
Allied Health: Nursing, Occupation or Physical Therapy, Nutrition, etc.	2.27%	97.73%
Earth and Environmental Sciences	4.55%	95.45%
Architecture, Fine and Visual Arts	0.00%	100.00%
History	0.00%	100.00%
English, Literature & Languages	0.00%	100.00%
Communications & Journalism	0.00%	100.00%
Engineering	2.70%	97.30%
Philosophy & Religion	0.00%	100.00%
Economics/Finance	5.56%	94.44%
Business (other than economics/finance)	1.54%	98.46%
Performing Arts	0.00%	100.00%
Psychology & Social Work	0.00%	100.00%
Medicine	0.00%	100.00%
Law	0.00%	100.00%
Mathematics/Statistics/Computer Science	0.00%	100.00%
Hard Sciences (Chemistry, Biology, Physics)	0.00%	100.00%
Other	0.00%	100.00%

Table 6.3.12 Have you personally ever taken part in an effort to develop any of the following? Highly detailed prompts for Gemini designed as a template to be used by others. Broken out by years at present institution

Years at present institution	Yes	No
3 years or less	0.00%	100.00%
3+ years to 5 years	0.00%	100.00%
5+ years to 10 years	2.60%	97.40%
10+ years to 20 years	1.99%	98.01%
More than 20 years	0.32%	99.68%

Table 6.4.1 Have you personally ever taken part in an effort to develop any of the following? A proprietary AI models.

	No Answer	Yes	No
Entire sample	0.00%	3.79%	96.21%

Table 6.4.2 Have you personally ever taken part in an effort to develop any of the following? A proprietary AI models. Broken out by enrollment

Enrollment	Yes	No
Less than 5,000	4.76%	95.24%
5,000+ - 17,000	2.40%	97.60%
17,000+ - 32,000	4.17%	95.83%
More than 32,000	3.82%	96.18%

Table 6.4.3 Have you personally ever taken part in an effort to develop any of the following? A proprietary AI models. Broken out by type of college or Carnegie Class

Type of college or Carnegie Class	Yes	No
Community College	2.70%	97.30%
BA-Granting Institution	7.27%	92.73%
Specialized College (i.e. performing arts, seminary, etc.)	5.56%	94.44%
MA-PHD Granting College	3.76%	96.24%
Level 1 or Level 2 Research University	3.36%	96.64%

Table 6.4.4 Have you personally ever taken part in an effort to develop any of the following? A proprietary AI models. Broken out by public or private college

Public or private college	Yes	No
Public	3.08%	96.92%
Private	5.34%	94.66%

Table 6.4.5 Have you personally ever taken part in an effort to develop any of the following? A proprietary AI models. Broken out by age of respondent

Age of respondent	Yes	No
30 or under	11.11%	88.89%
31-39	0.00%	100.00%
40-49	5.83%	94.17%
50-59	3.87%	96.13%
60 or over	3.46%	96.54%
Unknown	0.00%	100.00%

Table 6.4.6 Have you personally ever taken part in an effort to develop any of the following? A proprietary AI models. Broken out by political views

Political views	Yes	No
Left of Liberal	2.97%	97.03%
Liberal	3.14%	96.86%
Middle of the Road	2.07%	97.93%
Conservative	10.00%	90.00%
Unknown	0.00%	100.00%

Table 6.4.7 Have you personally ever taken part in an effort to develop any of the following? A proprietary AI models. Broken out by annual personal income

Annual personal income	Yes	No
Less than $50,000	0.00%	100.00%
$50,000 to $75,000	3.08%	96.92%
$75,000+ to $100,000	2.59%	97.41%
$100,000 to $150,000	3.98%	96.02%
More than $150,000	4.86%	95.14%
Unknown	0.00%	100.00%

Table 6.4.8 Have you personally ever taken part in an effort to develop any of the following? A proprietary AI models. Broken out by tenure track

Tenure track	Yes	No
Tenured	3.61%	96.39%
Not tenured but on a tenure track	5.41%	94.59%
Not tenured and not on a tenure track	3.90%	96.10%
Unknown	0.00%	100.00%

Table 6.4.9 Have you personally ever taken part in an effort to develop any of the following? A proprietary AI models. Broken out by gender of respondent

Gender of respondent	Yes	No
Male	2.94%	97.06%
Female	4.52%	95.48%
Transgender	0.00%	100.00%
Unknown	0.00%	100.00%

Table 6.4.10 Have you personally ever taken part in an effort to develop any of the following? A proprietary AI models. Broken out by race or ethnicity

Race or ethnicity	Yes	No
White or Caucasian	3.73%	96.27%
Black or African American	3.85%	96.15%
Hispanic or Latino	3.70%	96.30%
Asian or Asian American	2.63%	97.37%
Mixed Race	11.11%	88.89%
Unknown	0.00%	100.00%

Table 6.4.11 Have you personally ever taken part in an effort to develop any of the following? A proprietary AI models. Broken out by academic field

Academic field	Yes	No
Social Sciences (except economics and psychology)	4.62%	95.38%
Education or Library Science	0.00%	100.00%
Allied Health: Nursing, Occupation or Physical Therapy, Nutrition, etc.	0.00%	100.00%
Earth and Environmental Sciences	4.55%	95.45%
Architecture, Fine and Visual Arts	0.00%	100.00%
History	0.00%	100.00%
English, Literature & Languages	4.08%	95.92%
Communications & Journalism	18.18%	81.82%
Engineering	5.41%	94.59%
Philosophy & Religion	0.00%	100.00%
Economics/Finance	5.56%	94.44%
Business (other than economics/finance)	9.23%	90.77%
Performing Arts	0.00%	100.00%
Psychology & Social Work	0.00%	100.00%
Medicine	3.85%	96.15%
Law	0.00%	100.00%
Mathematics/Statistics/Computer Science	2.86%	97.14%
Hard Sciences (Chemistry, Biology, Physics)	3.88%	96.12%
Other	0.00%	100.00%

Table 6.4.12 Have you personally ever taken part in an effort to develop any of the following? A proprietary AI models. Broken out by years at present institution

Years at present institution	Yes	No
3 years or less	5.00%	95.00%
3+ years to 5 years	3.57%	96.43%
5+ years to 10 years	2.60%	97.40%
10+ years to 20 years	3.48%	96.52%
More than 20 years	4.14%	95.86%

Table 7.1 Have you ever sold or licensed any content developed by you to any developer of an AI model or engine?

	No Answer	Yes	No
Entire sample	9.39%	0.91%	89.70%

Table 7.2 Have you ever sold or licensed any content developed by you to any developer of an AI model or engine? Broken out by enrollment

Enrollment	No Answer	Yes	No
Less than 5,000	9.52%	0.60%	89.88%
5,000+ - 17,000	9.58%	0.60%	89.82%
17,000+ - 32,000	8.93%	1.19%	89.88%
More than 32,000	9.55%	1.27%	89.17%

Table 7.3 Have you ever sold or licensed any content developed by you to any developer of an AI model or engine? Broken out by type of college or Carnegie Class

Type of college or Carnegie Class	No Answer	Yes	No
Community College	8.11%	0.00%	91.89%
BA-Granting Institution	5.45%	0.00%	94.55%
Specialized College (i.e. performing arts, seminary, etc.)	5.56%	5.56%	88.89%
MA-PHD Granting College	9.02%	0.75%	90.23%
Level 1 or Level 2 Research University	10.31%	0.96%	88.73%

Table 7.4 Have you ever sold or licensed any content developed by you to any developer of an AI model or engine? Broken out by public or private college

Public or private college	No Answer	Yes	No
Public	10.57%	1.10%	88.33%
Private	6.80%	0.49%	92.72%

Table 7.5 Have you ever sold or licensed any content developed by you to any developer of an AI model or engine? Broken out by age of respondent

Age of respondent	No Answer	Yes	No
30 or under	11.11%	0.00%	88.89%
31-39	10.00%	0.00%	90.00%
40-49	7.77%	0.97%	91.26%
50-59	11.61%	1.94%	86.45%
60 or over	8.36%	0.58%	91.07%
Unknown	33.33%	0.00%	66.67%

Table 7.6 Have you ever sold or licensed any content developed by you to any developer of an AI model or engine? Broken out by political views

Political views	No Answer	Yes	No
Left of Liberal	7.92%	0.00%	92.08%
Liberal	9.80%	1.18%	89.02%
Middle of the Road	9.84%	0.52%	89.64%
Conservative	10.00%	2.00%	88.00%
Unknown	0.00%	0.00%	100.00%

Table 7.7 Have you ever sold or licensed any content developed by you to any developer of an AI model or engine? Broken out by annual personal income

Annual personal income	No Answer	Yes	No
Less than $50,000	18.75%	0.00%	81.25%
$50,000 to $75,000	4.62%	3.08%	92.31%
$75,000+ to $100,000	6.03%	0.86%	93.10%
$100,000 to $150,000	9.45%	0.50%	90.05%
More than $150,000	11.34%	0.81%	87.85%
Unknown	13.33%	0.00%	86.67%

Table 7.8 Have you ever sold or licensed any content developed by you to any developer of an AI model or engine? Broken out by tenure track

Tenure track	No Answer	Yes	No
Tenured	9.79%	1.03%	89.18%
Not tenured but on a tenure track	5.41%	0.00%	94.59%
Not tenured and not on a tenure track	8.66%	0.87%	90.48%
Unknown	50.00%	0.00%	50.00%

Table 7.9 Have you ever sold or licensed any content developed by you to any developer of an AI model or engine? Broken out by gender of respondent

Gender of respondent	No Answer	Yes	No
Male	10.66%	0.74%	88.60%
Female	8.51%	1.06%	90.43%
Transgender	20.00%	0.00%	80.00%
Unknown	0.00%	0.00%	100.00%

Table 7.10 Have you ever sold or licensed any content developed by you to any developer of an AI model or engine? Broken out by race or ethnicity

Race or ethnicity	No Answer	Yes	No
White or Caucasian	9.14%	0.75%	90.11%
Black or African American	11.54%	0.00%	88.46%
Hispanic or Latino	7.41%	0.00%	92.59%
Asian or Asian American	18.42%	0.00%	81.58%
Mixed Race	0.00%	11.11%	88.89%
Unknown	6.67%	0.00%	93.33%

Table 7.11 Have you ever sold or licensed any content developed by you to any developer of an AI model or engine? Broken out by academic field

Academic field	No Answer	Yes	No
Social Sciences (except economics and psychology)	4.62%	1.54%	93.85%
Education or Library Science	4.76%	2.38%	92.86%
Allied Health: Nursing, Occupation or Physical Therapy, Nutrition, etc.	13.64%	0.00%	86.36%
Earth and Environmental Sciences	4.55%	0.00%	95.45%
Architecture, Fine and Visual Arts	10.00%	0.00%	90.00%
History	0.00%	0.00%	100.00%
English, Literature & Languages	8.16%	2.04%	89.80%
Communications & Journalism	0.00%	0.00%	100.00%
Engineering	18.92%	0.00%	81.08%
Philosophy & Religion	16.67%	0.00%	83.33%
Economics/Finance	11.11%	0.00%	88.89%
Business (other than economics/finance)	13.85%	0.00%	86.15%
Performing Arts	0.00%	0.00%	100.00%
Psychology & Social Work	9.68%	3.23%	87.10%
Medicine	11.54%	1.28%	87.18%
Law	16.67%	0.00%	83.33%
Mathematics/Statistics/Computer Science	8.57%	0.00%	91.43%
Hard Sciences (Chemistry, Biology, Physics)	9.71%	0.00%	90.29%
Other	0.00%	25.00%	75.00%

Table 7.12 Have you ever sold or licensed any content developed by you to any developer of an AI model or engine? Broken out by years at present institution

Years at present institution	No Answer	Yes	No
3 years or less	15.00%	2.50%	82.50%
3+ years to 5 years	10.71%	3.57%	85.71%
5+ years to 10 years	3.90%	0.00%	96.10%
10+ years to 20 years	8.96%	1.00%	90.05%
More than 20 years	10.19%	0.64%	89.17%

To the best of your knowledge, has your college/university or any department or entity of it made any effort to collect and prepare content from its scholars to be used in an AI model of any kind? If so, what kind of efforts?

Social Sciences (except economics and psychology)

1 I know individual faculty have - no idea what the university plans to do.
2 No
3 No
4 Not to my knowledge
5 No
6 No.
7 No
8 I am not aware unfortunately
9 No and I do not think that the union contract or faculty senate would allow it.
10 Not to my knowledge
11 No idea. We're more concerned about students using AI to write papers.
12 Not that I know of.
13 No
14 No
15 Don't know.
16 Don't know. I retired in 2019, and the AI craze was not yet a big deal. My home situation was such that I quit nearly all academic work at retirement.
17 Not sure
18 I don't think so.
19 No awareness of this.
20 I don't know
21 Don't know
22 No idea
23 No
24 No idea, and the copyright concerns here are significant.
25 I do not know but I suspect yes.
26 No
27 I am not aware if this has been done.
28 They have provided some lectures and presentations.
29 Don't know of any such effort.

Economics/Finance

1 dedicated AI institute
2 No, focus has been on designing pedagogy around AI
3 Yes
4 Not aware
5 Not that I'm aware of.
6 No
7 No
8 Not that I am aware of

Business (other than economics/finance)

1 Yes
2 No knowledge
3 No
4 Not sure.
5 No
6 Yes, our university has developed its own internal AI model to be used as a resource for professors to embed tutorials into their classes
7 ACADEMIC DEPT HEAD CREATES PROMPT FOR USE BY DEPARTMENT
8 Uncertain
9 No effort that I know of.
10 Not that I know of
11 Am not aware of any such model.
12 No
13 No and none
14 No
15 Nothing to date.
16 No, they have not
17 Yes
18 Don't know
19 I don't know
20 No
21 No
22 No.
23 Not sure
24 No idea.
25 Not that I know of.
26 Not to my knowledge
27 Not sure
28 No.

Performing Arts

1 No
2 Not to my knowledge
3 Not sure
4 No
5 Not sure
6 No

Psychology & Social Work

1 Yes workshops
2 Not sure, I'm sure they have. There is a department dedicated to AI.
3 Not that I know of
4 Yes, there has been a concerted effort from our administration to encourage discussion and usage, through talks, discussion groups, book clubs to read books about AI, a symposium, etc.
5 No
6 Unsure
7 No. I don't believe this has happened at our institution
8 Not that I know of
9 No
10 Not sure

Medicine

1 No
2 Not known
3 No
4 Not sure
5 Unknown
6 No
7 I do not know
8 Aggregation of patient outcomes impact of specific types of imaging on changing patient management pulmonary modules. CT MR of brain quantification of brain beta amyloid and tau neurofibrillary tangles lines and wires on x-rays and scans pulmonary embolism.
9 No
10 No
11 No
12 No, not to my knowledge
13 Brain washing
14 Not that I am aware of
15 Yes of many kinds and sorts
16 I am not aware of any such effort.
17 Do not know

18 Yes, a GPT internal version is made.
19 No
20 No
21 don't know
22 No.
23 Don't know
24 No
25 Not sure
26 No
27 Unknown
28 No
29 Yes, our department has faculty who train computer programs to identify patterns/ cells in histology tissue sections. Our university has created a Center for AI.
30 No
31 No
32 Yes, development of Scholars content
33 No
34 I don't know
35 No
36 No knowledge
37 No
38 No

Law

1 Not sure.

Mathematics/Statistics/Computer Science

1 No
2 No
3 No
4 Not sure.
5 Its own AI programs for faculty and students to use.
6 No
7 Don't know so can't say.

Hard Sciences (Chemistry, Biology, Physics)

1 No
2 No
3 No
4 Not sure.
5 Its own AI programs for faculty and students to use.
6 No

7 Don't know so can't say.
1 No
2 No
3 No
4 No
5 Not sure
6 No
7 No
8 No idea
9 None that I know of.
10 No
11 Yes, but I do not familiar with this content.
12 No
13 don't know of any effort
14 No
15 I don't know.
16 No
17 don't know
18 Yes. Colleagues in Computer Science
19 No
20 Yes
21 No
22 No
23 No
24 No
25 No
26 I don't know.
27 Not sure about this
28 Do not know
29 I am not aware of any.
30 Bunch of projects going on in the Data Science Institute (Nirav Merchant e.g.). The cancer center, where I am at, is trying to gather data from patient records for use in creating impactful clinical trials in metastatic setting. Work in digital twin.
31 No
32 Don't know

Education or Library Science

1 Not significant for now
2 None known
3 No
4 NO.
5 I don't know
6 A number of committees with a variety of AI focuses
7 No

8 Emails requesting joining forces to prepare content to be used in classes and scholarly paper preparation
9 Not to my knowledge -because I haven't searched AI for that, I don't know
10 No
11 I don't know.
12 No
13 Not sure
14 That's an excellent question. I believe the university is starting such efforts but I have not participated. I am a great believer in tech (particularly assistive tech for people with disabilities) but I am not jumping on the AI bandwagon until the field has some serious discussions about the ethics of AI scraping, the lack of oversight, the privacy issues, and illustrations and text being used without permission from the original content provider.
15 No
16 Not as far as I know. To the best of my knowledge our university is not yet providing AI for faculty use. We buy it on our own.
17 Purchase of AI tools, creating policies about AI
18 No
19 I don 't know

Allied Health: Nursing, Occupation or Physical Therapy, Nutrition, etc.

1 I am not aware
2 No
3 I am not aware
4 No
5 No
6 No
7 CSU has a system wide "EDU Workspace" with ChatGPT Open AI where all engagement is used to train a model only for the CSU system
8 Yes
9 Unsure
10 Unknown
11 No
12 Not to my knowledge
13 Not aware
14 Not that I am aware of occurring.

Earth and Environmental Sciences

1 Not that I know of.
2 Not aware of any efforts.
3 Yes, I ignore it.
4 no idea, no desire to know
5 I don't know, but I think not.

6 Don't know
7 I am involved with the policy making on AI usage in my department and university. We, as committee members, have made enormous efforts to facilitate the use of AI.

Architecture, Fine and Visual Arts

1 Not sure
2 Do not know; have seen no calls for this.
3 I have not heard of any such efforts. I would not contribute.
4 Not that I know of.
5 I'm not sure. I haven't heard.

History

1 Not that I'm aware of.
2 No
3 Don't know.
4 No.
5 Not sure
6 I do not know.
7 No
8 I don't know.

English, Literature & Languages

1 No
2 I have no idea. Why would anyone do this?
3 Not that I know of, and I hope not
4 Responsible AI use in research and writing: Guidelines emphasize the ethical use of AI, data privacy and intellectual property considerations, and proper attribution for AI-assisted work. Integrating AI into teaching and learning: Universities are exploring using AI for tasks like personalized learning, automated grading, and generating educational materials, according to QuadC. Addressing academic integrity concerns: Policies are being developed to define acceptable AI use in student assignments and research, with some institutions prohibiting AI-generated content or requiring disclosure.
5 Not sure
6 Not that I know.
7 My department now does mostly in-class writing without access to any electronic tools to assess students' thinking and writing abilities.
8 Not sure if this answers your question, but the UT System Academy has helped interested faculty develop AI teaching assistants for our classes
9 No
10 Not sure but yes.

11 No
12 No idea
13 NONE in my department. If the University is doing it, they don't promote it well.
14 Not that I know of and I wouldn't contribute
15 Not sure, figure efforts on campus are there though.
16 Yes, custom chatbots for individual courses
17 No
18 I do not know. No information has been provided to me about such efforts if they are ongoing.
19 I have no idea, but would not be surprised if they were.

Communications & Journalism

1 No, not to my knowledge.
2 Yes.
3 Not that I know of.
4 No.
5 Yes
6 Not that I know of.

Engineering

1 No
2 None so far.
3 Not at all. We're way behind.
4 No, I am unaware of any university effort toward collecting or preparing content from its scholars for use in an AI model.
5 Not aware of any
6 No
7 Don't know
8 No
9 No
10 I do not know of any effort
11 No

Philosophy & Religion

1 No
2 Yes. Unknown
3 Yes. And they are clearly using AI to generate news reports.

www.ingramcontent.com/pod-product-compliance
Ingram Content Group UK Ltd.
Pitfield, Milton Keynes, MK11 3LW, UK
UKHW051508290926
15471UKWH00008B/1493

9 798885 173087